STUDENT'S SOLUTIONS MANUAL

NANCY S. BOUDREAU

A SECOND COURSE IN STATISTICS

REGRESSION ANALYSIS

FIFTH EDITION

WILLIAM MENDENHALL
TERRY SINCICH

PRENTICE HALL, Upper Saddle River, NJ 07458

Production Editor: *James Buckley*
Supplement Acquisitions Editor: *Ann Heath*
Production Coordinator: *Joan Eurell*
Manufacturing Buyer: *Ben Smith*

Simon & Schuster / A Viacom Company
Upper Saddle River, New Jersey 07458

Printed in the United States of America

10 9 8 7 6 5 4 3 2 1

ISBN 0-13-456468-5

Prentice-Hall International (UK) Limited, *London*
Prentice-Hall of Australia Pty. Limited, *Sydney*
Prentice-Hall Canada Inc., *Toronto*
Prentice-Hall Hispanoamericana, S.A., *Mexico*
Prentice-Hall of India Private Limited, *New Delhi*
Prentice-Hall of Japan, Inc., *Tokyo*
Simon & Schuster Asia Pte. Ltd., *Singapore*
Editora Prentice-Hall do Brasil, Ltda., *Rio de Janeiro*

Preface

This solutions manual is designed to accompany the text *A Second Course in Business Statistics: Regression Analysis*, Fifth Edition. It provides answers to most odd-numbered exercises for each chapter in the text. Other methods of solution may also be appropriate; however, the author has presented one that she believes to be most instructive to the statistics student. The student should first attempt to solve the assigned exercises without help from this manual. Then, if unsuccessful, the solution in the manual will clarify points necessary to the solution. The student who successfully solves an exercise should still refer to the manual's solution. Many points are clarified and expanded upon to provide maximum insight into and benefit from each exercise.

Instructors will also benefit from the use of this manual. It will save time in preparing presentations of the solutions and possibly provide another point of view regarding their meaning.

Some of the exercises are subjective in nature and thus omitted from the Answer Key at the end of *A Second Course in Business Statistics: Regression Analysis,* Fifth Edition. The subjective decisions regarding these exercises have been made and are explained by the author. Solutions based on these decisions are presented; the solution to this type of exercise is often most instructive. When an alternative interpretation of an exercise may occur, the author has often addressed it and given justification for the approach taken.

I would like to thank Brenda Dobson and Kelly Evans for typing the manuscript and creating the art work.

Nancy S. Boudreau
Bowling Green State University
Bowling Green, Ohio

Contents

1 A Review of Basic Concepts (Optional)

1.1 a. The experimental units are the new automobiles. The model name, manufacturer, type of transmission, engine size, number of cylinders, estimated city miles/gallon, and estimated highway miles are measured on each automobile.

b. Model name, manufacturer, and type of transmission are **qualitative**. None of these is measured on a numerical scale. Engine size, number of cylinders, estimated city miles/gallon, and estimated highway miles/gallon are all **quantitative**. Each of these variables is measured on a numerical scale.

1.3

1. Pipe diameter is a **quantitative** variable since it is measured on a numerical scale (inches).
2. Pipe material is a **qualitative** variable since it produces nonnumerical data (copper, iron, etc.).
3. Age is a **quantitative** variable since it is measured on a numerical scale (year).
4. Location is a **qualitative** variable since it produces nonnumerical data.
5. Pipe length is a **quantitative** variable since it is measured on a numerical scale (inches, feet, etc.).
6. Stability of surrounding soil is a **qualitative** variable since it produces nonnumerical data (unstable, moderately stable, or stable).
7. Corrosiveness of surrounding soil is a **qualitative** variable since it produces nonnumerical data (corrosive or noncorrosive).
8. Internal pressure is a **quantitative** variable since it is measured on a numerical scale (pounds per square inch.)
9. Percentage of pipe covered with land cover is a **quantitative** variable since it is measured on a numerical scale.
10. Breakage rate is a **quantitative** variable since it is measured on a numerical scale (number of times pipe had to be repaired due to breakage).

1.5 a. Chip discharge rate (number of chips discarded per minute) is quantitative. The number of chips is a numerical value.

b. Drilling depth (millimeters) is quantitative. The depth is a numerical value.

c. Oil velocity (millimeters per second) is quantitative. The velocity is a numerical value.

d. Type of drilling (single-edge, BTA, or ejector) is qualitative. The type of drilling is not a numerical value.

e. Quality of hole surface is qualitative. The quality can be judged as poor, good, excellent, etc., which are categories and are not numerical values.

1.7 a. Since the variable **city** is measured on a nonnumerical scale, the variable is qualitative. Examples are New Orleans, New York City, etc.

b. Since the variable **region of the county** is measured on a nonnumerical scale, the variable is qualitative. Examples are northwest region, central region, etc.

c. Since the variable **home type** is measured on a nonnumerical scale, the variable is qualitative. Examples are starter, trade-up, etc.

d. Since the variable **number of days on the market** is measured on a numerical scale, the variable is quantitative. Examples are 60 days, 45 days, etc.

e. Since the variable **sale price** is measured on a numerical scale, the variable is quantitative. Examples are $109,000, $250,000, etc.

1.9 a. The population of interest is the thion levels of all possible daily ambient air specimens that can be collected at the orchard.

b. The sample of interest is the thion levels for the 13 ambient air specimens actually measured at the orchard.

1.11 a. The experimental units are the smokers and former smokers.

b. There were three variables measured on each subject: type of smoker (current or former), gender (male or female), and weight gain (slight, moderate, significant, or major). All three of these variables are qualitative. All are measured on a nonnumerical scale.

c. The data presented in the exercise represent a sample. Data were collected on only 1,885 current smokers and 768 former smokers.

1.13 a. The population of interest is the job status (quit or not) of all state lottery winners who win big payoffs.

b. The sample of interest is the job status of the 576 lottery winners who responded to the questionnaire.

c. The inference in this situation is that 11% of all state lottery winners who win big payoffs quit their jobs during the first year after striking it rich.

1.15 a. The population of interest to the Concern for Dying is the opinions of all American adults concerning the support of euthanasia.

b. No. There are over 100 million American adults. It would be almost impossible to survey all of them.

c. The results of the survey could possibly be used to develop legislation regarding euthanasia.

1.17 a. The graph is a frequency histogram with chess ratings on the horizontal axis and frequency on the vertical axis.

b. There appears to be about 150 players with ratings between 2400 and 2500. There appears to be about 75 players with ratings between 2500 and 2600. There appears to be about 50 players with ratings between 2600 and 2700. Thus, there appears to be about 125 + 75 + 50 = 250 players rated as grand masters.

c. The data are skewed to the right. The tail to the right of the modal class (1300 to 1400) is longer than the tail to the left of the modal class.

1.19 a. To convert a frequency histogram to a relative frequency histogram, we must first divide each of the frequencies by the sum of all the frequencies, which is 1 + 1 + 3 + 4 + 10 + 7 + 5 + 4 + 3 + 2 + 3 + 2 + 3 + 2 = 50. The relative frequency table is:

Class Interval	Frequency	Relative Frequency
5.5–15.5	1	1/50 = .02
15.5–25.5	1	1/50 = .02
25.5–35.5	3	3/50 = .06
35.5–45.5	4	4/50 = .08
45.5–55.5	10	10/50 = .20
55.5–65.5	7	7/50 = .14
65.5–75.5	5	5/50 = .10
75.5–85.5	4	4/50 = .08
85.5–95.5	3	3/50 = .06
95.5–105.5	2	2/50 = .04
145.5–155.5	3	3/50 = .06
165.5–175.5	2	2/50 = .04
175.5–185.5	3	3/50 = .06
195.5–205.5	2	2/50 = .04

The relative frequency histogram is:

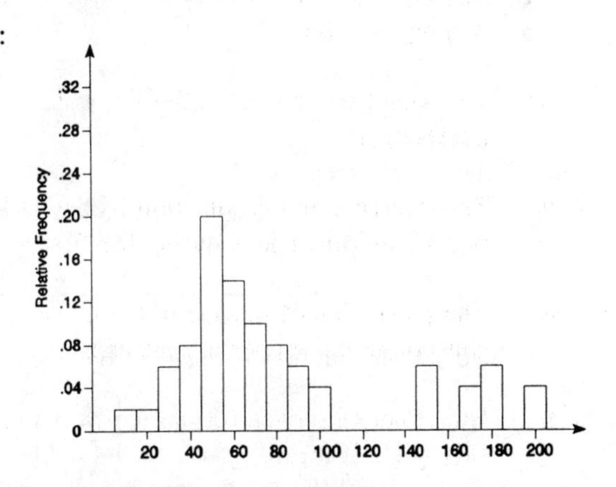

b. It would be very unusual to observe a drill chip with a length of at least 190 mm. There are only 2 out of 50 drill chips that are 190 mm or longer. The proportion of drill chips with lengths of at least 190 mm is .04.

1.21 a. The data represent a sample. These data represent only 62 of the major colleges and universities in the U.S.

b. We will use a relative frequency histogram to describe the data. Suppose we select 8 intervals. The range of the data is 26.7 − 3.7 = 23.0. The class interval width is the range divided by the number of intervals and is 23/8 ≈ 2.9. To ensure that no data point falls on a class boundary, we will choose 3.65 for the lower limit of the first interval. The upper limit of the first interval is 3.65 + 2.9 = 6.55. This process is continued until we have our eight intervals.

Class	Class Interval	Tally	Class Frequency	Class Relative Frequency
1	3.65 - 6.55	||||	4	.067
2	6.55 - 9.45	~~||||~~ ~~||||~~ ||||	14	.233
3	9.45 - 12.35	~~||||~~ ~~||||~~ ~~||||~~ |||	18	.300
4	12.35 - 15.25	~~||||~~ ~~||||~~	10	.167
5	15.25 - 18.15	~~||||~~ ||	7	.117
6	18.15 - 21.05	|||	3	.050
7	21.05 - 23.95			1	.017
8	23.95 - 26.85	|||	3	.050
		Totals	60	1.001

The relative frequency histogram is:

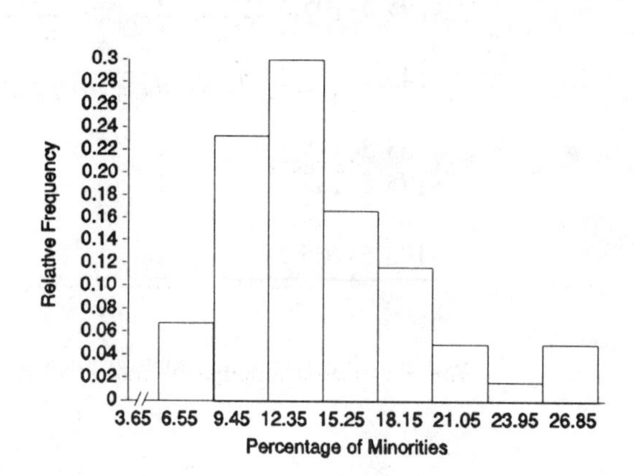

1.23 a. The sample mean is:

$$\bar{y} = \frac{\sum_{i=1}^{n} y_i}{n} = \frac{1 + 5 + 0 + 2 + 5 + 7 + 1}{7} = \frac{21}{7} = 3$$

The sample variance is obtained by first calculating

$$\sum_{i=1}^{n} (y_i - \bar{y})^2 = \sum y_i^2 - \frac{\left[\sum_{i=1}^{n} y_i\right]^2}{n}$$

$$= (1^2 + 5^2 + 0^2 + 2^2 + 5^2 + 7^2 + 1^2) - \frac{(21)^2}{7} = 105 - 63 = 42$$

Then, $$s^2 = \frac{\sum_{i=1}^{n}(y_i - \bar{y})^2}{n - 1} = \frac{42}{7 - 1} = 7$$

$$s = \sqrt{s^2} = \sqrt{7} = 2.646$$

b. The sample mean is:

$$\bar{y} = \frac{\sum_{i=1}^{n} y_i}{n} = \frac{1 + 2 + 0 + 0 + 5 + 4}{6} = \frac{12}{6} = 2$$

The sample variance is obtained by first calculating

$$\sum_{i=1}^{n}(y_i - \bar{y})^2 = \sum_{i=1}^{n} y_i^2 - \frac{\left[\sum_{i=1}^{n} y_i\right]^2}{n}$$

$$= (1^2 + 2^2 + 0^2 + 0^2 + 5^2 + 4^2) - \frac{(12)^2}{6} = 46 - 24 = 22$$

Then, $$s^2 = \frac{\sum_{i=1}^{n}(y_i - \bar{y})^2}{n - 1} = \frac{22}{6 - 1} = 4.4$$

$$s = \sqrt{s^2} = \sqrt{4.4} = 2.098$$

c. The sample mean is:

$$\bar{y} = \frac{\sum_{i=1}^{n} y_i}{n} = \frac{10 + 8 + 12 + 2}{4} = \frac{32}{4} = 8$$

The sample variance is obtained by first calculating

$$\sum_{i=1}^{n}(y_i - \bar{y})^2 = \sum_{i=1}^{n} y_i^2 - \frac{\left[\sum_{i=1}^{n} y_i\right]^2}{n}$$

$$= (10^2 + 8^2 + 12^2 + 2^2) - \frac{(32)^2}{4} = 312 - 256 = 56$$

Then, $$s^2 = \frac{\sum_{i=1}^{n}(y_i - \bar{y})^2}{n - 1} = \frac{56}{4 - 1} = 18.6667$$

$$s = \sqrt{s^2} = \sqrt{18.6667} = 4.320$$

d. The sample mean is:

$$\bar{y} = \frac{\sum_{i=1}^{n} y_i}{n} = \frac{3 + 4 + 10 + 2}{4} = \frac{19}{4} = 4.75$$

The sample variance is obtained by first calculating

$$\sum_{i=1}^{n} (y_i - \bar{y})^2 = \sum_{i=1}^{n} y_i^2 - \frac{\left(\sum_{i=1}^{n} y_i\right)^2}{n}$$

$$= (3^2 + 4^2 + 10^2 + 2^2) - \frac{(19)^2}{4} = 129 - 90.25 = 38.75$$

Then, $$s^2 = \frac{\sum_{i=1}^{n} (y_i - \bar{y})^2}{n - 1} = \frac{38.75}{4 - 1} = 12.9167$$

$$s = \sqrt{s^2} = \sqrt{12.9167} = 3.594$$

e. The sample mean is:

$$\bar{y} = \frac{\sum_{i=1}^{n} y_i}{n} = \frac{1 + 1 + 20 + 20 + 8}{5} = \frac{50}{5} = 10$$

The sample variance is obtained by first calculating

$$\sum_{i=1}^{n} (y_i - \bar{y})^2 = \sum_{i=1}^{n} y_i^2 - \frac{\left(\sum_{i=1}^{n} y_i\right)^2}{n}$$

$$= (1^2 + 1^2 + 20^2 + 20^2 + 8^2) - \frac{(50)^2}{5} = 866 - 500 = 366$$

Then, $$s^2 = \frac{\sum_{i=1}^{n} (y_i - \bar{y})^2}{n - 1} = \frac{366}{5 - 1} = 91.5$$

$$s = \sqrt{s^2} = \sqrt{91.5} = 9.566$$

f. $$\bar{y} = \frac{\sum_{i=1}^{n} y_i}{n} = \frac{2 + 100 + 104 + 2}{4} = \frac{208}{4} = 52$$

$$s^2 = \frac{\sum_{i=1}^{n} y_i^2 - \frac{\left(\sum_{i=1}^{n} y_i\right)^2}{n}}{n-1} = \frac{2^2 + 100^2 + 104^2 + 2^2 - \frac{(208)^2}{4}}{4-1}$$

$$= \frac{20824 - 10816}{3} = \frac{10008}{3} = 3{,}336$$

$$s = \sqrt{s^2} = \sqrt{3336} = 57.758$$

g. $$\bar{y} = \frac{\sum_{i=1}^{n} y_i}{n} = \frac{(-1) + (-3) + (-2) + 0 + (-3) + (-3)}{6} = \frac{-12}{6} = -2$$

$$s^2 = \frac{\sum_{i=1}^{n} y_i^2 - \frac{\left(\sum_{i=1}^{n} y_i\right)^2}{n}}{n-1} = \frac{(-1)^2 + (-3)^2 + (-2)^2 + 0^2 + (-3)^2 + (-3)^2 - \frac{(-12)^2}{6}}{6-1}$$

$$= \frac{32 - 24}{5} = \frac{8}{5} = 1.6$$

$$s = \sqrt{s^2} = \sqrt{1.6} = 1.265$$

h. $$\bar{y} = \frac{\sum_{i=1}^{n} y_i}{n} = \frac{\frac{1}{5} + \frac{1}{5} + \frac{1}{5} + \frac{2}{5} + .2 + \frac{4}{5}}{6} = \frac{2}{6} = .333$$

$$s^2 = \frac{\sum_{i=1}^{n} y_i^2 - \frac{\left(\sum_{i=1}^{n} y_i\right)^2}{n}}{n-1}$$

$$= \frac{\left[\left(\frac{1}{5}\right)^2 + \left(\frac{1}{5}\right)^2 + \left(\frac{1}{5}\right)^2 + \left(\frac{2}{5}\right)^2 + .2^2 + \left(\frac{4}{5}\right)^2\right] - \frac{(2)^2}{6}}{6-1}$$

$$= \frac{.96 - .6667}{5} = \frac{.2933}{5} = .0587$$

$$s = \sqrt{s^2} = \sqrt{.0587} = .242$$

1.25 a. Using Guideline #2, we know that about 95% of the pressure readings will fall within 2 standard deviations of the mean, i.e., within the interval:

$$\bar{y} \pm 2s \Rightarrow 7.99 \pm 2(2.02) \Rightarrow 7.99 \pm 4.04 \Rightarrow (3.95, 12.03)$$

b. No. Since about 95% of the pressure readings will fall between 3.95 and 12.03, it would be extremely unlikely to observe a pressure reading of 20 psi.

1.27 If the data set is of moderate size, we would expect approximately 95% of the measurements to lie within 2 standard deviations of the mean. Thus, we would expect the range of the data set to be approximately 4 standard deviations:

$$\begin{aligned} \text{range} &\approx 4s \\ 900 - 50 &\approx 4s \\ 850 &\approx 4s \\ s &\approx 212.5 \end{aligned}$$

1.29 a. The sample mean is:

$$\bar{y} = \frac{\sum_{i=1}^{n} y_i}{n} = \frac{5 + 10 + 3 + \cdots + 6}{30} = \frac{280}{30} = 9.333$$

The sample variance is obtained by first calculating

$$\sum_{i=1}^{n} (y_i - \bar{y})^2 = \sum_{i=1}^{n} y_i^2 - \frac{\left[\sum_{i=1}^{n} y_i\right]^2}{n} = (5^2 + 10^2 + 3^2 + \cdots + 6^2) - \frac{(280)^2}{30}$$

$$= 3{,}556 - 2{,}613.3333 = 942.6667$$

Then, $$s^2 = \frac{\sum_{i=1}^{n} (y_i - \bar{y})^2}{n - 1} = \frac{942.6667}{30 - 1} = 32.5057$$

$$s = \sqrt{s^2} = \sqrt{32.5057} = 5.701$$

b. Most of the scores should fall within 2 standard deviations of the mean. This interval would be:

$$\bar{y} \pm 2s \Rightarrow 9.333 \pm 2(5.701) \Rightarrow 9.333 \pm 11.402 \Rightarrow (-2.069, 20.735)$$

1.31 a. The mean default rate is 14.682.

b. The variance is 199.974 and the standard deviation is 14.141.

c. Using Guideline #2, we would expect about .95 of the measurements to fall within two standard deviations of the mean.

d. The interval in part (c) would be:

$$\bar{y} \pm 2s \Rightarrow 14.682 \pm 2(14.141) \Rightarrow 14.682 \pm 28.282 \Rightarrow (-13.600, 42.964)$$

Thus, 60 of the 66 observations or 60/66 = .909 of the observations fall within 2 standard deviations of the mean. This is a little less than expected.

e. If the largest data point was eliminated from the data set, the mean would decrease. Similarly, if the largest data point was eliminated, the variability or standard deviation would decrease.

f. The mean is $\bar{y} = \frac{\sum_{i=1}^{n} y_i}{n} = \frac{892.8}{65} = 13.735$

The new mean (13.735) is less than the original mean (14.682).

The variance is:

$$s^2 = \frac{\sum_{i=1}^{n} y_i^2 - \frac{\left[\sum_{i=1}^{n} y_i\right]^2}{n}}{n - 1} = \frac{21418.54 - \frac{892.8^2}{65}}{65 - 1} = \frac{9155.58862}{64} = 143.056$$

The standard deviation is $s = \sqrt{143.056} = 11.960$. Again, this new standard deviation is less than the original standard deviation (14.141).

g. Using Guideline #2, we would expect about .95 of the measurements to fall within two standard deviations of the mean.

The new interval would be:

$$\bar{y} \pm 2s \Rightarrow 13.735 \pm 2(11.96) \Rightarrow 13.735 \pm 23.92 \Rightarrow (-10.185, 37.655)$$

Thus, 59 of the 65 observations or $59/65 = .908$ of the observations fall within 2 standard deviations of the mean. This is again a little less than expected.

1.33 a. The probability that a normal random variable will lie between 1 standard deviation below the mean and 1 standard deviation above the mean is indicated by the shaded area in the figure:

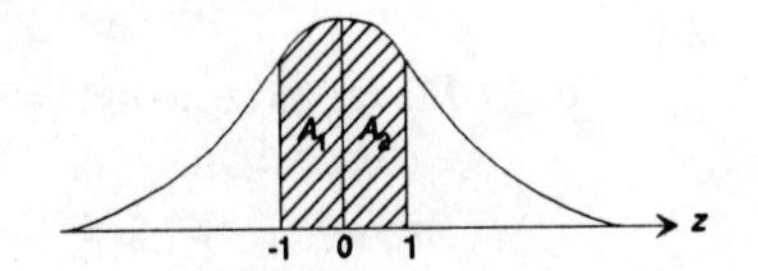

The desired probability is:

$$P(-1 \le z \le 1) = P(-1 \le z \le 0) + P(0 \le z \le 1) = A_1 + A_2.$$

Now, $A_1 = P(-1 \le z \le 0)$
$= P(0 \le z \le 1)$ (by symmetry of the normal distribution)
$= .3413$ (from Table 1)

and $A_2 = P(0 \le z \le 1)$
$= .3413$ (from Table 1)

Thus, $P(-1 \le z \le 1) = .3413 + .3413 = .6826$

b. $P(-1.96 \le z \le 1.96) = P(-1.96 \le z \le 0) + P(0 \le z \le 1.96)$
$= P(0 \le z \le 1.96) + P(0 \le z \le 1.96)$
$= .4750 + .4750 = .9500$

c. $P(-1.645 \le z \le 1.645) = P(-1.645 \le z \le 0) + P(0 \le z \le 1.645)$
$= P(0 \le z \le 1.645) + P(0 \le z \le 1.645)$

$$\text{Now, } P(0 \le z \le 1.645) = \frac{P(0 \le z \le 1.64) + P(0 \le z \le 1.65)}{2}$$
$$= \frac{.4495 + .4505}{2} = .4500$$

Thus, $P(-1.645 \le z \le 1.645) = .4500 + .4500 = .9000$

d. $P(-3 \le z \le 3) = P(-3 \le z \le 0) + P(0 \le z \le 3)$
$= P(0 \le z \le 3) + P(0 \le z \le 3)$
$= .4987 + .4987 = .9974$

1.35 a. The first step is to determine whether the value of z_0 should be positive or negative. Since we want $P(z \le z_0) = .0013$, and we know that $P(z \le 0) = .5$, then it can be concluded that $z_0 < 0$.

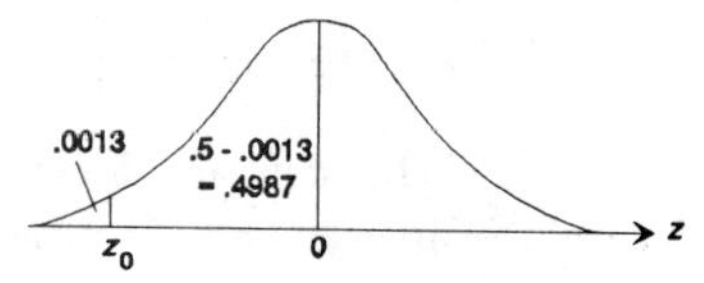

Thus, $P(z_0 \le z \le 0) = .5 - .0013 = .4987$ and from Table 1, $z_0 = -3$.

b. Notice, $P(-z_0 \le z \le z_0) = .95$ can be rewritten as:

$$P(-z_0 \le z \le 0) + P(0 \le z \le z_0) = .95$$
$$P(0 \le z \le z_0) + P(0 \le z \le z_0) = .95$$
$$2P(0 \le z \le z_0) = .95$$
$$P(0 \le z \le z_0) = .4750$$

From Table 1, $z_0 = 1.96$ and $-z_0 = -1.96$.

c. $P(-z_0 \le z \le z_0) = .90$

$$P(-z_0 \le z \le 0) + P(0 \le z \le z_0) = .90$$
$$P(0 \le z \le z_0) + P(0 \le z \le z_0) = .90$$
$$2P(0 \le z \le z_0) = .90$$
$$P(0 \le z \le z_0) = .4500$$

This value appears midway between the entries .4495 and .4505 in the body of Table 1. The location of this value implies that $z_0 = 1.645$ and $-z_0 = -1.645$.

d. $P(-z_0 \le z \le z_0) = .6826$

$$P(-z_0 \le z \le 0) + P(0 \le z \le z_0) = .6826$$
$$P(0 \le z \le z_0) + P(0 \le z \le z_0) = .6826$$
$$2P(0 \le z \le z_0) = .6826$$
$$P(0 \le z \le z_0) = .3413$$

From Table 1, $z_0 = 1$ and $-z_0 = -1$.

e. $P(-z_0 \le z \le 0) = .0596$
$P(0 \le z \le z_0) = .0596$

From Table 1, $z_0 = .15$ and $-z_0 = -.15$.

1.37 a. Let y = ingestion time $\qquad z = \frac{y - \mu}{\sigma} = \frac{4 - 2.83}{.79} = 1.48$

$P(y \geq 4) = P(z \geq 1.48) = .5 - P(0 \leq z \leq 1.48) = .5 - .4306 = .0694$

b. $z = \frac{y - \mu}{\sigma} = \frac{2 - 2.83}{.79} = -1.05 \qquad z = \frac{y - \mu}{\sigma} = \frac{3 - 2.83}{.79} = .22$

$P(2 \leq y \leq 3) = P(-1.05 \leq z \leq .22) = .3531 + .0871 = .4402$

1.39 a. A part is out of spec if it is less than .304 or greater than .322. Thus,

$$\begin{aligned} P(y < .304) + P(y > .322) &= P\left[z < \frac{.304 - .3015}{.0016}\right] + P\left[z > \frac{.322 - .3015}{.0016}\right] \\ &= P(z < 1.56) + P(z > 12.81) \\ &= (.5 + .4406) + (.5 - .5) \\ &= .9406 \end{aligned}$$

b. A part is out of spec if it is less than .304 or greater than .322. Thus,

$$\begin{aligned} P(y < .304) + P(y > .322) &= P\left[z < \frac{.304 - .3146}{.0030}\right] + P\left[z > \frac{.322 - .3146}{.0030}\right] \\ &= P(z < -3.53) + P(z > 2.47) \\ &= (.5 - .5) + (.5 - .4932) \\ &= .0068 \end{aligned}$$

This probability is much smaller than the probability in part (a). It appears that the changes made did improve product quality.

1.41 a. The probability a randomly selected birder will have a satisfactory score of at least 5 is represented by the shaded area:

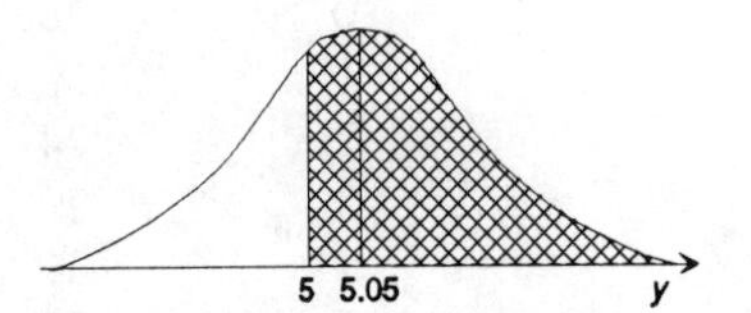

The corresponding z-value for the score of 5 is:

$$z = \frac{y - \mu}{\sigma} = \frac{5 - 5.05}{.98} = \frac{-.05}{.98} = -.05$$

Then, the area to the right of $y = 5$ is the same as the area to the right of $z = -.05$, a standard normal random variable.

$P(y > 5) = P(z > -.05)$

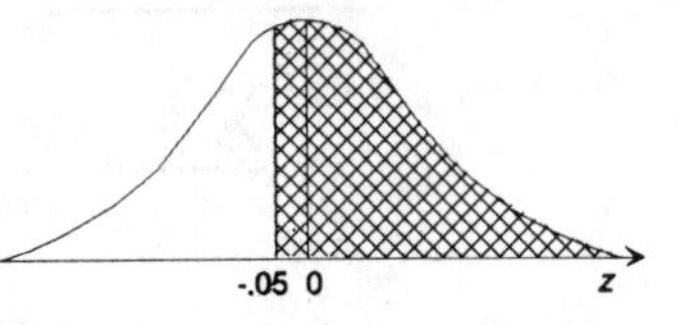

We now use Table 1 of Appendix C to find the probability between $z = 0$ and $z = -.05$. We find this area to be .0199. The area to the right of $z = -.05$ is then $.0199 + .5 = .5199$.

b. We want to find the point y_0, so that $P(y < y_0) = .25$.

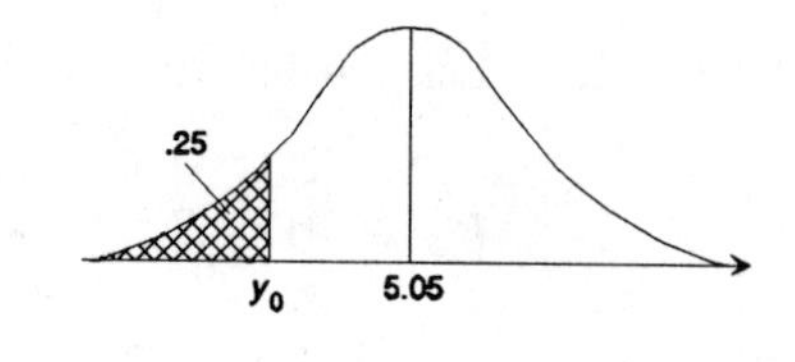

The corresponding z-value for y_0 is:

$$z = \frac{y - \mu}{\sigma} = \frac{y_0 - 5.05}{.98}$$

Using the standard normal curve, our picture becomes:

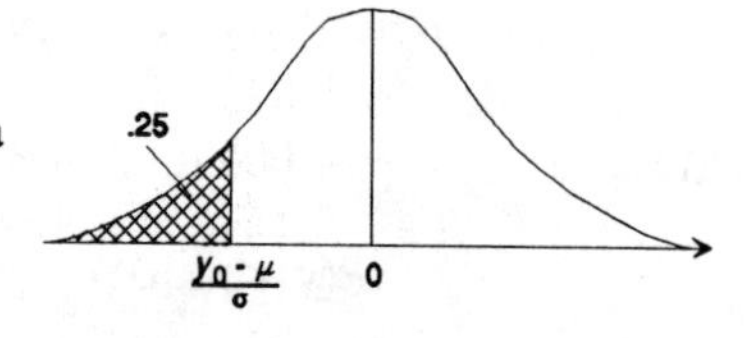

Since a total of .5 lies to the left of 0, we know the area between $z = 0$ and $z = \frac{y_0 - \mu}{\sigma}$ is $.5 - .25 = .25$.

Using Table 1, we can look up .25 as a probability to find the z-score at which it occurs. The closest z-score to the probability .25 is $z = .67$.

Since $\frac{y_0 - 5.05}{.98} < 0$, we know the z-score we want is $-.67$.

Now solve $\quad -.67 = \frac{y_0 - 5.05}{.98}$

$$(.98)(-.67) = y_0 - 5.05$$

$$y_0 = 5.05 + (.98)(-.67) = 4.3934$$

1.43 a. The relative frequency distribution is:

Value	Frequency	Relative Frequency
0	26	26/300 = .087
1	30	30/300 = .100
2	24	.080
3	29	.097
4	31	.103
5	25	.083
6	42	.140
7	36	.120
8	27	.090
9	30	.100
	300	1.000

b. $$\bar{y} = \frac{\sum y_i}{n} = \frac{1404}{300} = 4.68$$

c. $$s^2 = \frac{\sum y_i^2 - \frac{\left(\sum y_i\right)^2}{n}}{n - 1} = \frac{8942 - \frac{1404^2}{300}}{300 - 1} = 7.9307$$

d. The 50 sample means are:

4.833	4.500	4.500	5.667
4.667	5.000	4.167	5.000
5.167	4.667	5.333	4.167
4.500	5.333	3.833	2.500
5.667	3.833	4.333	2.667
5.000	4.167	4.833	5.500
7.333	4.000	3.500	2.167
5.833	3.333	3.500	7.000
4.000	4.333	6.833	5.833
6.167	4.000	6.833	2.667
3.167	3.833	5.833	5.667
4.833	5.167	3.833	5.500
5.500	3.500		

The frequency distribution for $\bar{y}$ is:

Sample Mean	Frequency	Relative Frequency
2.000 - 2.999	4	4/50 = .08
3.000 - 3.999	9	9/50 = .18
4.000 - 4.999	16	.32
5.000 - 5.999	16	.32
6.000 - 6.999	3	.06
7.000 - 7.999	2	.04
	50	1.00

The mean of the sample means is:

$$\bar{\bar{y}} = \frac{\sum \bar{y}_i}{n} = \frac{234}{50} = 4.68$$

$$s_{\bar{y}}^2 = \frac{\sum \bar{y}^2 - \frac{(\sum \bar{y})^2}{n}}{n - 1} = \frac{1162.483337 - \frac{234^2}{50}}{50 - 1} = 1.375$$

1.45 a. Since n is fairly large, the sampling distribution of $\bar{y}$ is approximately normal (by the Central Limit Theorem). The mean of the sampling distribution is:

$$\mu_{\bar{y}} = \mu = 15$$

The standard deviation of the sampling distribution is:

$$\sigma_{\bar{y}} = \frac{\sigma}{\sqrt{n}} = \frac{5}{\sqrt{20}} = 1.118$$

b. $P(\bar{y} \le 6) = P\left[z \le \frac{6 - 15}{1.118}\right] = P(z \le -8.05) \approx .5 - .5 \approx 0$ (using Table 1, Appendix C)

c. Since it would be almost impossible to observe a sample mean of 6 or smaller if there was no effect due to the hormone (probability $\approx$ 0), we would conclude that those taking the melatonin fell asleep quicker than those not taking the hormone. Thus, it appears that melatonin is a sleep-inducing hormone.

1.47 a. The point estimate of μ is $\bar{y} = 5.4$ hours.

b. For confidence coefficient $.99 = 1 - \alpha \Rightarrow \alpha = 1 - .99 = .01$. $\alpha/2 = .01/2 = .005$. From Table 1, Appendix C, $z_{.005} \approx 2.58$. The 99% confidence interval is:

$$\bar{y} \pm z_{\alpha/2}\frac{s}{\sqrt{n}} \Rightarrow 5.4 \pm 2.58\frac{3.6}{\sqrt{73}} \Rightarrow 5.4 \pm 1.09 \Rightarrow (4.31,\ 6.49)$$

c. We can be 99% confident that the interval (4.31, 6.49) encloses the true mean number of hours spent by a reviewer in conducting a complete review of a paper submitted to AMJ or AMR.

1.49 a. For confidence coefficient $.98 = 1 - \alpha \Rightarrow \alpha = 1 - .98 = .02$; $\alpha/2 = .02/2 = .01$. From Table 2, Appendix C, with df $= n - 1 = 13 - 1 = 12$, $t_{.01} = 2.681$. The small sample confidence interval for μ is:

$$\bar{y} \pm t_{.01}\left(\frac{s}{\sqrt{n}}\right) \Rightarrow 74.31 \pm 2.681\left(\frac{20.94}{\sqrt{13}}\right) \Rightarrow 74.31 \pm 15.570 \Rightarrow (58.740,\ 89.880)$$

b. Yes. Since $\mu = 100$ is not contained in the 98% confidence interval, it is not a likely value for the true value of μ. We are 98% confident the true average collision-damage rating of all station wagons and mini-vans is between 58.740 and 89.880.

1.51 First, compute the following:

$$\bar{y} = \frac{\sum y}{n} = \frac{2.5 + 3.1 + \cdots + 4.1}{20} = \frac{119.8}{20} = 5.99$$

$$s^2 = \frac{\sum y^2 - \frac{(\sum y)^2}{n}}{n - 1} = \frac{1972.82 - \frac{(119.8)^2}{20}}{20 - 1} = 66.064$$

$$s = \sqrt{66.064} = 8.128$$

a. For confidence coefficient $.90 = 1 - \alpha \Rightarrow \alpha = 1 - .90 = .10$; $\alpha/2 = .10/2 = .05$. From Table 2, Appendix C, $t_{.05} = 1.729$ with 19 df. The 90% confidence interval is:

$$\bar{y} \pm t_{\alpha/2}\left(\frac{s}{\sqrt{n}}\right) \Rightarrow 5.99 \pm 1.729\left(\frac{8.128}{\sqrt{20}}\right) \Rightarrow 5.99 \pm 3.142 \Rightarrow (2.848,\ 9.132)$$

b. We are 90% confident the true mean TCDD level in plasma of all Vietnam veterans exposed to Agent Orange is between 2.848 and 9.132 parts per trillion.

c. We must assume that the TCDD levels in plasma of all Vietnam veterans exposed to Agent Orange are normally distributed.

d. A stem-and-leaf display of the data is as follows:

Stem	Leaves
1.	6, 8, 8
2.	0, 1, 5, 5
3.	0, 1, 1, 3, 5
4.	1, 6, 7
5.	
6.	0, 9
7.	2
⋮	
20.	0
⋮	
36.	0

From the stem-and-leaf display, the data are skewed to the right. The data are not normally distributed.

1.53 a. The confidence coefficient is .95.

b. In repeated sampling, 95% of the intervals created would contain the population mean rate.

c. We are 95% confident that the population mean rate is contained in the interval 49.66 to 51.48.

d. We must assume that the population rates of the passive samplers are approximately normally distributed.

1.55 a. For confidence coefficient $.98 = 1 - \alpha \Rightarrow \alpha = 1 - .98 = .02$; $\alpha/2 = .02/2 = .01$. From Table 1, Appendix C, $z_{.01} = 2.33$. The 98% confidence interval is:

$$\bar{y} \pm z_{\alpha/2}\frac{\sigma}{\sqrt{n}} \Rightarrow 1.94 \pm 2.33\left(\frac{.92}{\sqrt{290}}\right) \Rightarrow 1.94 \pm .126 \Rightarrow (1.814, 2.066)$$

We are 98% confident the mean attitudinal score of marketing faculty is between 1.814 and 2.066.

b. We could reduce the width of the confidence interval by decreasing the level of confidence or by increasing the sample size.

1.57 a. The rejection region is determined by the sampling distribution of the test statistic, the direction of the test ($>$, $<$, or $\neq$), and the tester's choice of α.

b. No, nothing is **proven**. When the decision based on sample information is to reject H_0, we run the risk of committing a Type I error. We might have decided in favor of the research hypothesis when, in fact, the null hypothesis was the true statement. The existence of Type I and Type II errors makes it impossible to **prove** anything using sample information.

1.59 a. To determine if the average number of faxes transmitted in the United States each minute is 88,000, we test:

H_0: $\mu = 88{,}000$
H_a: $\mu \neq 88{,}000$

b. First, we would have to determine the necessary sample size. Once the sample size was determined, we would select a random sample of one-minute intervals and count the number of faxes transmitted for each interval. From this sample, we would calculate the sample average number of faxes transmitted per minute and use this to test the above hypotheses.

1.61 We want to test:

H_0: $\mu = 2$
H_a: $\mu > 2$

where μ is the true mean number of suppliers engaged by farm and power equipment dealers.

Rejection region: At $\alpha = .05$, the rejection region for this one-tailed test is:

$z > z_\alpha$
or $z > z_{.05}$
or $z > 1.645$ (Table 1, Appendix C)

Test statistic: $$z = \frac{\bar{y} - \mu_0}{\sigma_{\bar{y}}} = \frac{\bar{y} - \mu_0}{\sigma/\sqrt{n}} \approx \frac{\bar{y} - \mu_0}{s/\sqrt{n}} = \frac{3.12 - 2}{1.91/\sqrt{226}} = 8.82$$

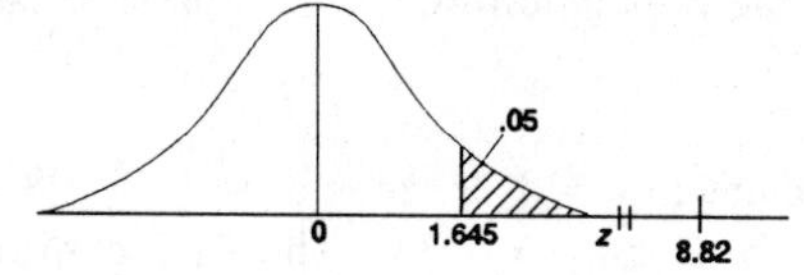

Conclusion: Reject H_0. There is sufficient evidence (at $\alpha = .05$) to conclude that the mean number of suppliers engaged by farm and power equipment dealers exceeds 2. The p-value for this problem is p-value $= P(z > 8.82) \approx .5 - .5 = 0$.

1.63 We want to test:

H_0: $\mu = 4.0$
H_a: $\mu < 4.0$

where μ is the true mean radium-226 level of soil specimens.

Test statistic: $$t = \frac{\bar{y} - \mu_0}{s/\sqrt{n}} = \frac{2.413 - 4}{2.081/\sqrt{26}} = -3.89$$

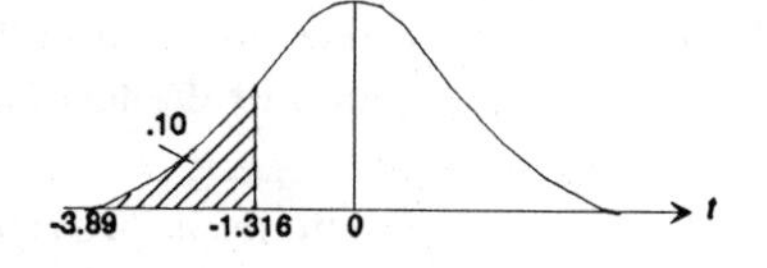

Rejection region: $\alpha = .10$, df $= n - 1 = 25$
$t_{.10} = 1.316$.

Reject H_0 if $t < -1.316$

Conclusion: Reject H_0. There is sufficient evidence (at $\alpha = .10$) to indicate the mean radium-226 level of soil specimens in southern Dade County is less than the EPA limit of 4.0 pCi/L.

Assumption: The distribution of radium-226 level of soil specimens is approximately normal.

1.65 H_0: $\mu = 1920$
H_a: $\mu < 1920$

where μ is the true mean throughput per 40-hour week for the system.

Test statistic: $t = \dfrac{\bar{y} - \mu_0}{s/\sqrt{n}} = \dfrac{1908.8 - 1920}{18/\sqrt{5}} = -1.39$

Rejection region: $\alpha = .05$, df $= n - 1 = 4$,
$t_\alpha = t_{.05} = 2.132$.

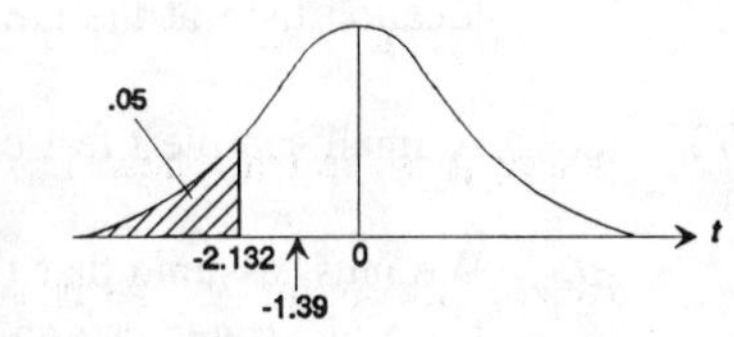

Reject H_0 if $t < -2.132$

Conclusion: Do not reject H_0. There is insufficient evidence (at $\alpha = .05$) to indicate that the mean throughput per 40-hour week for the system is less than 1920 parts.

Assumption: The distribution of throughputs per 40-hour week for the system is approximately normal.

1.67 H_0: $\mu = .91$
H_a: $\mu < .91$

where μ is the true mean digestibility coefficient for crude protein in cats.

Test statistic: $t = \dfrac{\bar{y} - \mu_0}{s/\sqrt{n}} = \dfrac{.81 - .91}{.042/\sqrt{28}} = -12.60$

Rejection region: $\alpha = .01$, df $= n - 1 = 27$,
$t_{.01} = 2.473$.

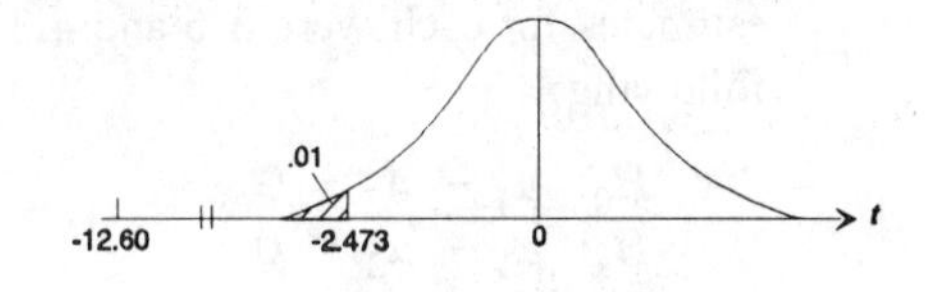

Reject H_0 if $t < -2.473$

Conclusion: Reject H_0. There is sufficient evidence (at $\alpha = .01$) to indicate that the mean digestibility coefficient for crude protein in cats is less than .91.

Assumption: The distribution of the digestibility coefficients for crude protein in cats is approximately normal.

1.69 The two populations must have:

1) relative frequency distributions that are approximately normal, and
2) variances σ_1^2 and σ_2^2 that are equal.

The two samples must both have been randomly and independently chosen.

1.71 a. The general form of a large sample 90% confidence interval for $(\mu_1 - \mu_2)$ is:

$$(\bar{y}_1 - \bar{y}_2) \pm z_{.05}\sqrt{\frac{\sigma_1^2}{n_1} + \frac{\sigma_2^2}{n_2}}$$

Using s_1^2 and s_2^2 to estimate σ_1^2 and σ_2^2, and substituting the values of the sample statistics yields:

$$(9.80 - 9.42) \pm 1.645\sqrt{\frac{.5409^2}{30} + \frac{.4789^2}{30}} \Rightarrow .38 \pm .217 \Rightarrow (.163, .597)$$

b. No. Since the interval constructed in part (a) contains only positive values, we can conclude that there is evidence that the mean voltage readings are higher at the old location than at the new location.

1.73 a. A small sample t test can be used to compare the means.

b. We must assume that the students were randomly and independently assigned to the two teaching strategies and that the scores of the two groups are normally distributed.

c. The statistics 3.28 and 3.40 are only point estimates of their corresponding population means. Without any information concerning the sampling distribution of $\bar{y}_1 - \bar{y}_2$, we cannot perform a test of hypothesis.

d. Since $p = .79$ is very high, we must fail to reject H_0. For any reasonable choice of α, there is insufficient evidence to indicate the mean scores for the use of clear/sterile gloves questions differ for the two teaching strategies.

e. Since $p = .02$ is very low, we can reject H_0 for any $\alpha > p = .02$. For $\alpha = .025$, there is sufficient evidence to indicate the mean scores for the choice of stethoscope question do differ for the two teaching strategies.

1.75 Let μ_1 = mean awareness score in 1989 and μ_2 = mean awareness score in 1990. The point estimates for each were 3.6 and 4.3, respectively. The test of hypothesis that was run is the following:

H_0: $\mu_1 - \mu_2 = 0$
H_a: $\mu_1 - \mu_2 < 0$

Test statistic: $$z \approx \frac{(\bar{y}_1 - \bar{y}_2) - D_0}{\sqrt{\frac{s_1^2}{n_1} + \frac{s_2^2}{n_2}}} = \frac{(3.6 - 4.3) - 0}{\sqrt{\frac{1.26^2}{1211} + \frac{1.24^2}{2006}}} = -15.36$$

The p-value for this test is $p = P(z \leq -15.36) \approx .5 - .5 = 0$.

This is consistent with what the researchers reported.

1.77 a. The small sample confidence interval for $\mu_1 - \mu_2$ is:

$$(\bar{y}_1 - \bar{y}_2) \pm t_{\alpha/2}\sqrt{\frac{s_1^2}{n_1} + \frac{s_2^2}{n_2}}$$

where $t_{\alpha/2}$ is based on $n_1 + n_2 - 2 = 10 + 10 - 2 = 18$ df.

Confidence coefficient $.90 \Rightarrow 1 - \alpha \Rightarrow \alpha = 1 - .90 = .10$ and $\alpha/2 = .10/2 = .05$. From Table 2, $t_{.05} = 1.734$ with 18 df. The 90% confidence interval is:

$$(3.35 - 2.36) \pm 1.734\sqrt{\frac{(.79)^2}{10} + \frac{(.47)^2}{10}} \Rightarrow .99 \pm .50 \Rightarrow (.49, 1.49)$$

b. Yes, since the interval constructed in part (a) contains only positive values, we are 90% confident that it takes longer, on average, for the urchins to ingest the green blades than the decayed blades.

1.79 H_0: $\mu_1 - \mu_2 = 0$
H_a: $\mu_1 - \mu_2 \neq 0$

where μ_1 = average investment/quad for plants using electrical utilities and μ_2 = average investment/quad for plants using gas utilities.

From the printout, the test statistic is $t = .68$ and the p-value is $p = .50$.

Conclusion: Do not reject H_0. There insufficient evidence (at $\alpha = .05$) to conclude that there is a difference in average investment/quad between plants using electrical utilities and those using gas utilities.

Assumptions: This procedure requires the assumption that the samples of investment/quad values are randomly and independently selected from normal populations with equal variances.

1.81 H_0: $\frac{\sigma_1^2}{\sigma_2^2} = 1$ vs. H_a: $\frac{\sigma_1^2}{\sigma_2^2} \neq 1$

where σ_1^2 and σ_2^2 are the population variances of ME content in cats fed canned food and dry food, respectively.

Test statistic: $F = \frac{\text{larger sample variance}}{\text{smaller sample variance}} = \frac{s_2^2}{s_1^2} = \frac{(.48)^2}{(.26)^2} = 3.41$

Rejection region: $\alpha = .10$, $n_2 - 1 = 28$ numerator degrees of freedom, $n_1 - 1 = 27$ denominator degrees of freedom, $F_{.05} \approx 1.88$.

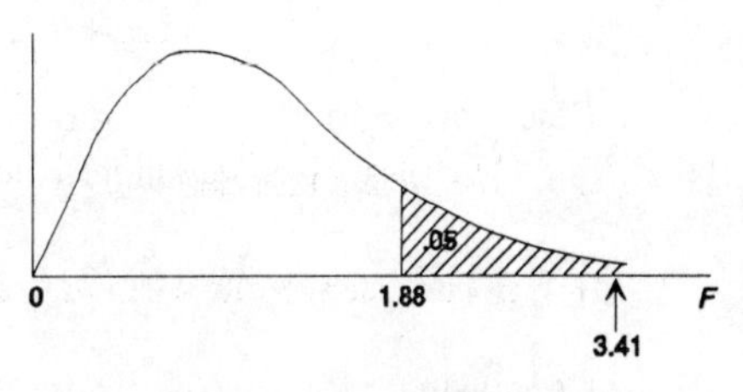

Reject H_0 if $F > 1.88$

Conclusion: Reject H_0. There is sufficient evidence (at $\alpha = .10$) to conclude that the variation of ME content in cats fed canned food differs from the variation of ME content in cats fed dry food.

Assumptions: The distributions of ME content in cats fed canned and dry food are both approximately normal. The samples of ME content are randomly and independently selected from their populations.

1.83 H_0: $\dfrac{\sigma_1^2}{\sigma_2^2} = 1$ vs. H_a: $\dfrac{\sigma_1^2}{\sigma_2^2} \neq 1$

where σ_1^2 and σ_2^2 are the population variances of log-price ratios of rural and urban counties, respectively.

Test statistic: $F = \dfrac{\text{larger sample variance}}{\text{smaller sample variance}} = \dfrac{s_1^2}{s_2^2} = \dfrac{.310^2}{.199^2} = 2.43$

Rejection region: Using $\alpha = .05$, $n_1 - 1 = 9$ numerator degrees of freedom, $n_2 - 1 = 22$ denominator degrees of freedom, $F_{.025} = 2.76$.

Reject H_0 if $F > 2.76$

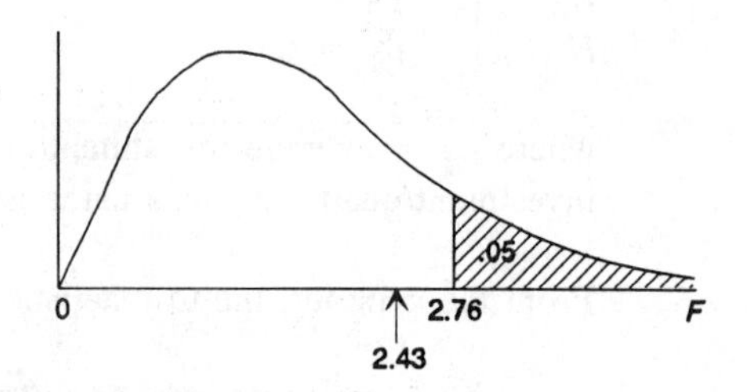

Conclusion: Do not reject H_0. There is insufficient evidence (at $\alpha = .05$) to conclude that the variances in the log-price ratios of rural and urban counties differ.

Assumptions: The distributions of the log-price ratios for rural and urban counties are both approximately normal. The samples of the log-price ratios are randomly and independently selected from their populations.

1.85 H_0: $\sigma_1^2/\sigma_2^2 = 1$ $(\sigma_1^2 \neq \sigma_2^2)$

H_a: $\sigma_1^2/\sigma_2^2 \neq 1$ $(\sigma_1^2 \neq \sigma_2^2)$

where σ_1^2 = variance of the one wet sampler readings and σ_2^2 = variance of the three wet sampler readings.

Test statistic: $F = \dfrac{\text{larger sample variance}}{\text{smaller sample variance}} = \dfrac{s_1^2}{s_2^2} = \dfrac{6.3^2}{2.6^2} = 5.871$

Rejection region: Using $\alpha = .05$, $n_1 - 1 = 364$ numerator df, $n_2 - 1 = 364$ denominator df, $F_{.025} \approx 1.00$.

Reject H_0 if $F > 1.00$

Conclusion: Reject H_0. There is sufficient evidence (at $\alpha = .05$) to indicate the variations in hydrogen readings for the two sampling schemes differ.

Assumptions: The distributions of the hydrogen readings for the one wet sampler and the three wet samplers are both approximately normal. The samplers of the hydrogen readings are randomly and independently selected from their populations.

1.87 a. $$\bar{y} = \frac{\sum_{i=1}^{n} y_i}{n} = \frac{11 + 2 + 2 + 1 + 9}{5} = \frac{25}{5} = 5$$

$$s^2 = \frac{\sum_{i=1}^{n} (y_i - \bar{y})^2}{n - 1} = \frac{\sum_{i=1}^{n} y_i^2 - \frac{\left(\sum_{i=1}^{n} y_i\right)^2}{n}}{n - 1} = \frac{(11^2 + 2^2 + 2^2 + 1^2 + 9^2) - \frac{(25)^2}{5}}{5 - 1}$$

$$= \frac{211 - 125}{4} = \frac{86}{4} = 21.5$$

$$s = \sqrt{s^2} = \sqrt{21.5} \approx 4.637$$

b. $$\bar{y} = \frac{\sum_{i=1}^{n} y_i}{n} = \frac{22 + 9 + 21 + 15}{4} = \frac{67}{4} = 16.75$$

$$s^2 = \frac{\sum_{i=1}^{n} (y_i - \bar{y})^2}{n - 1} = \frac{\sum_{i=1}^{n} y_i^2 - \frac{\left(\sum_{i=1}^{n} y_i\right)^2}{n}}{n - 1} = \frac{(22^2 + 9^2 + 21^2 + 15^2) - \frac{(67)^2}{4}}{4 - 1}$$

$$= \frac{1231 - 1122.25}{3} = \frac{108.75}{3} = 36.25$$

$$s = \sqrt{s^2} = \sqrt{36.25} \approx 6.021$$

c. $$\bar{y} = \frac{\sum_{i=1}^{n} y_i}{n} = \frac{34}{7} = 4.857$$

$$s^2 = \frac{\sum_{i=1}^{n} y_i^2 - \frac{\left(\sum_{i=1}^{n} y_i\right)^2}{n}}{n - 1} = \frac{344 - \frac{(34)^2}{7}}{7 - 1} = \frac{178.857}{6} \approx 29.81$$

$$s = \sqrt{s^2} = \sqrt{29.81} \approx 5.460$$

d. $$\bar{y} = \frac{\sum_{i=1}^{n} y_i}{n} = \frac{16}{4} = 4$$

$$s^2 = \frac{\sum_{i=1}^{n} y_i^2 - \frac{\left[\sum_{i=1}^{n} y_i\right]^2}{n}}{n - 1} = \frac{64 - \frac{(16)^2}{4}}{4 - 1} = \frac{0}{3} \approx 0$$

$$s = \sqrt{s^2} = \sqrt{0} = 0$$

1.89 a. $z = \frac{y - \mu}{\sigma} = \frac{10 - 30}{5} = \frac{-20}{5} = -4$

The sign and magnitude of the z-value indicate that the y-value is 4 standard deviations below the mean.

b. $z = \frac{y - \mu}{\sigma} = \frac{32.5 - 30}{5} = \frac{2.5}{5} = .5$

The y-value is .5 standard deviation above the mean.

c. $z = \frac{y - \mu}{\sigma} = \frac{30 - 30}{5} = \frac{0}{5} = 0$

The y-value is equal to the mean of the random variable y.

d. $z = \frac{y - \mu}{\sigma} = \frac{60 - 30}{5} = \frac{30}{5} = 6$

The y-value is 6 standard deviations above the mean.

1.91 The stem-and-leaf display of the effect size index for the brand name indicates that the index ranges from .03 to .60. The stem-and-leaf display of the effect size index for the store name indicates that the index ranges from .00 to .44. Also, .667 (10 out of 15) of the brand name indexes are larger than .09 while only .176 (3 out of 17) of the store name indexes are larger than .09. Thus, it appears that the brand name has a stronger effect on perceived quality.

1.93 a. To find the value z_0 such that .5 units of probability lie above z_0, we remember that $P(z \geq 0) = .5$ for the standard normal distribution. Thus, $z_0 = 0$.

b. Since $P(z \leq z_0) = .5199$, it must be true that z_0 is positive. Then $P(0 \leq z \leq z_0) = .5199 - .5 = .0199$, and the location of .0199 in the body of Table 1 implies $z_0 = .05$.

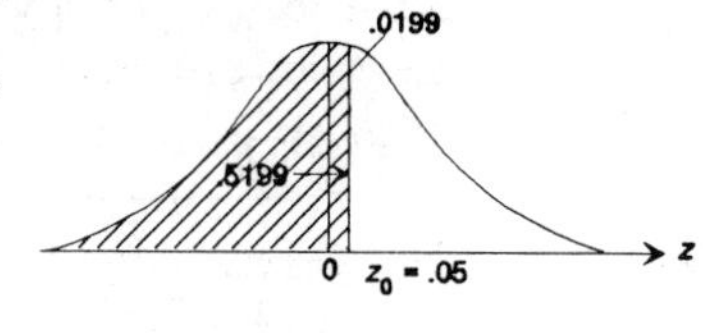

c. In order to have $P(z \geq z_0) = .3300$, the value of z_0 must be positive. Then $P(0 \leq z \leq z_0) = .5 - .3300 = .1700$, and the location of .1700 in the body of Table 1 implies $z_0 = .44$.

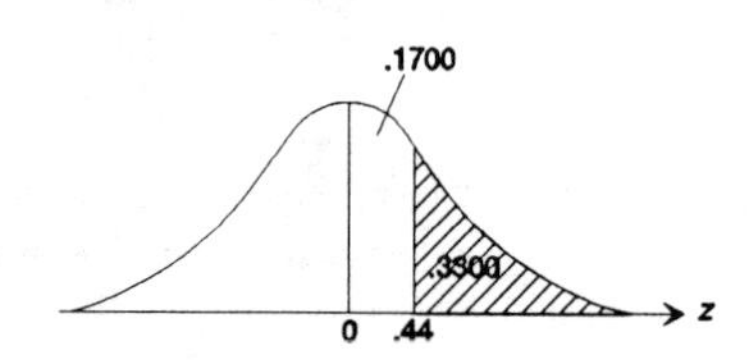

d. In order to have $P(z_0 \leq z \leq .59) = .5845$, the value of z_0 must be negative.

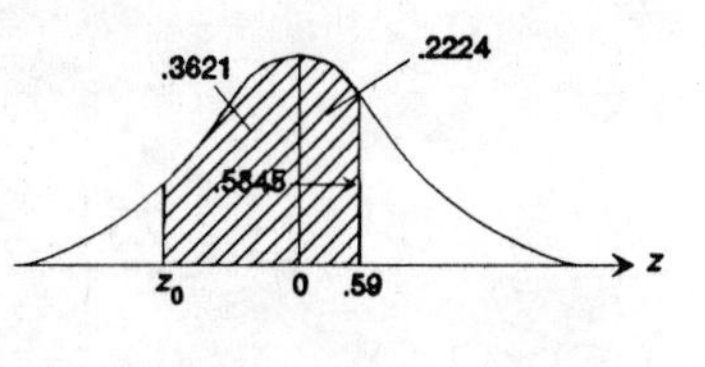

Then $P(z_0 \leq z \leq 0) = .5845 - .2224 = .3621$, and the location of .3621 in the body of Table 1 implies $z_0 = -1.09$.

1.95 Let y = length of life for calf implanted with an artificial heart. For y, $\mu = 80$, $\sigma = 25$, and y is approximately normally distributed.

a. $$P(y > 120) = P\left[z > \frac{120 - 80}{25}\right]$$
$$= P(z > 1.6)$$
$$= .5 - P(0 \leq z \leq 1.6)$$
$$= .5 - .4452 = .0548$$

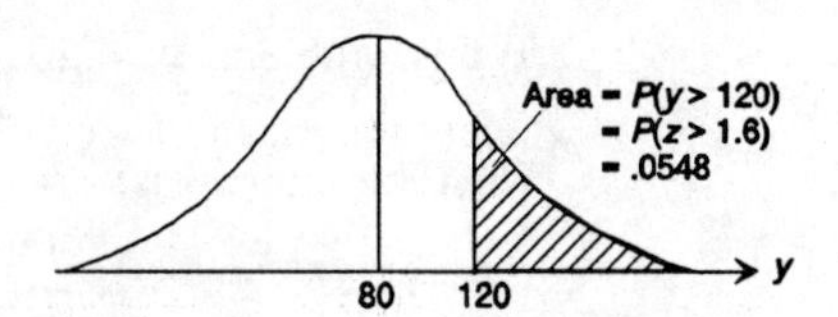

b. Since $P(y > d) = .25$ is required in the upper tail, there must be $.5 - .25 = .2500$ between d and 80.

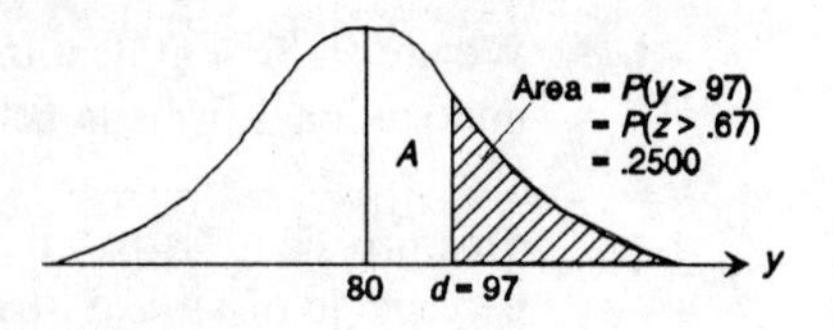

$A = .5 - .25 = .2500$
$z_0 = .67$
$d = \mu + z_0\sigma = 80 + .67(25) = 96.75 \approx 97$

1.97 a. Let y = the diameter at breast height of white spruce trees.

$$P(y < 12) = P\left[z < \frac{12 - \mu}{\sigma}\right] = P\left[z < \frac{12 - 17}{6}\right]$$
$$= P(z < -.83)$$
$$= .5 - .2967$$
$$= .2033$$

b. Since the probability of observing a white spruce tree with breast diameter less than 12 meters is .2033, which is not small, it would not be an unusual event.

c. $$P(y > 37) = P\left[z > \frac{37 - 17}{6}\right] = P(z > 3.33) \approx .5 - .5 = 0$$

d. Since the probability of observing a white spruce with breast diameter of more than 37 meters is essentially 0, it would be very unusual to observe a white spruce with a breast diameter of 38 meters.

1.99 a. The general form of a small sample 90% confidence interval for a population mean is:

$$\bar{y} \pm t_{.05}\frac{s}{\sqrt{n}}$$

where t is based on $n - 1$ degrees of freedom. It is necessary to compute $\bar{y}$ and s, the sample mean and sample standard deviation, respectively:

$$\bar{y} = \frac{\sum_{i=1}^{n} y_i}{n} = \frac{73.2 + 58.9 + 26.6 + \cdots + 34.6 + 20.5}{20} = \frac{822.7}{20} = 41.135$$

$$s^2 = \frac{\sum_{i=1}^{n} y_i^2 - \frac{\left[\sum_{i=1}^{n} y_i\right]^2}{n}}{n-1} = \frac{\left[(73.2)^2 + (58.9)^2 + \cdots + (20.5)^2\right] - \frac{(822.7)^2}{20}}{20-1}$$

$$= \frac{38685.13 - 33841.7645}{19} = 254.914$$

$$s = \sqrt{s^2} = \sqrt{254.914} = 15.966$$

In this problem, $n = 20$, so $t_{.05}$ (19 df) $= 1.729$

Then the interval is:

$$41.135 \pm 1.729 \frac{15.966}{\sqrt{20}} \Rightarrow 41.135 \pm 6.193 \Rightarrow (34.962, 47.308)$$

b. We are 90% confident that the mean foreign revenue percentage of all U.S. based multinational firms is between 34.962% and 47.308%.

c. This procedure is based on the assumption that the distribution of foreign revenue percentage of all U.S.-based multinational firms is approximately normal.

1.101 a. By the Central Limit Theorem, the sampling distribution of $\bar{y}_{37}$ is approximately normally distributed with mean $\mu_{\bar{y}_{37}} = \mu = 8$ and a standard deviation $\sigma_{\bar{y}_{37}} = \sigma/\sqrt{n} = 5/\sqrt{37} = .82199$.

b. The z-score for $\bar{y} = 10$ is $z = \frac{\bar{y} - \mu_{\bar{y}}}{\sigma_{\bar{y}}} = \frac{10 - 8}{.82199} = 2.43$

$P(\bar{y} > 10) = P(z > 2.43) = .5 - .4925 = .0075$

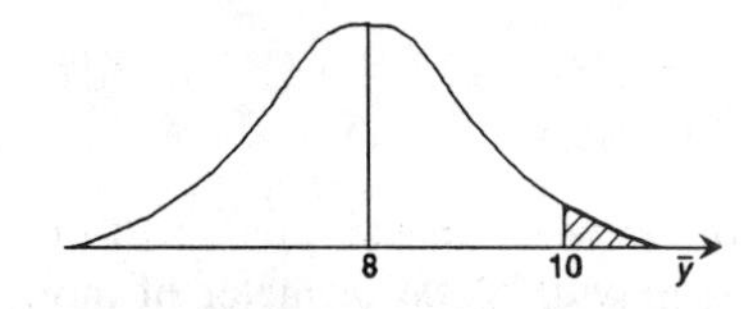

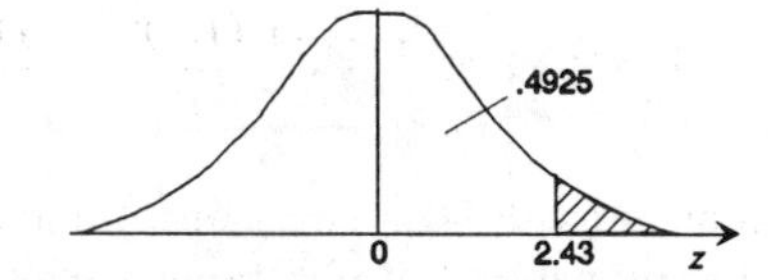

c. We expect $\bar{y}$ to fall within $\mu_{\bar{y}} \pm 2\sigma_{\bar{y}} \Rightarrow 8 \pm 2(.822) \Rightarrow 8 \pm 1.64 = (6.36, 9.64)$

1.103 $H_0\colon \frac{\sigma_1^2}{\sigma_2^2} = 1$ vs. $H_a\colon \frac{\sigma_1^2}{\sigma_2^2} \neq 1$

where σ_1^2 and σ_2^2 are the population variances of the test scores of on-campus and off-campus students, respectively.

Test statistic: $F = \frac{\text{larger sample variance}}{\text{smaller sample variance}} = \frac{s_1^2}{s_2^2} = \frac{2.86^2}{1.42^2} = 4.06$

Rejection region: $\alpha = .10$, $n_1 - 1 = 14$ numerator degrees of freedom, $n_2 - 1 = 14$ denominator degrees of freedom, $F_{.05} \approx 2.47$.

Reject H_0 if $F > 2.47$

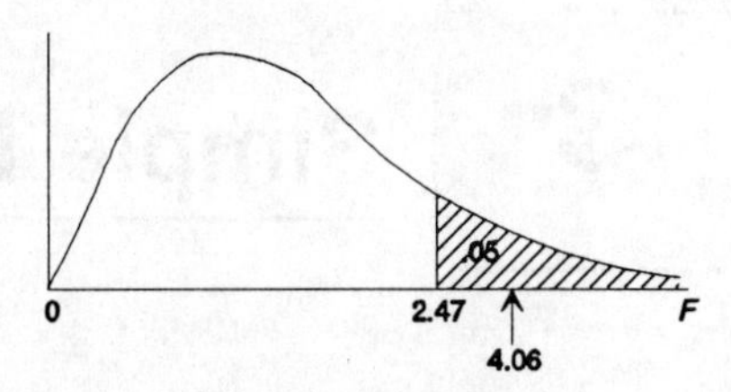

Conclusion: Reject H_0. There is sufficient evidence (at $\alpha = .10$) to conclude that the population variances are significantly different.

Assumptions: The distributions of the test scores of on-campus and off-campus are both approximately normal. The samples of the test scores are randomly and independently selected from their populations.

This result is significant as it implies that the assumption of equal variances necessary when using the t distribution to make inferences about $(\mu_1 - \mu_2)$ is probably not valid. This assumption is necessary for the completion of Exercise 1.78. Thus, the results of Exercise 1.78 are questionable.

1.105 a. Let μ_1 = mean number of promotions for female managers at large firms and μ_2 = mean number of promotions for female managers at small firms.

The point estimate for $\mu_1 - \mu_2$ is $\bar{y}_1 - \bar{y}_2 = 1.0 - .9 = .1$

b. The general form of a large sample 90% confidence interval for $(\mu_1 - \mu_2)$ is:

$$(\bar{y}_1 - \bar{y}_2) \pm z_{.05}\sqrt{\frac{\sigma_1^2}{n_1} + \frac{\sigma_2^2}{n_2}}$$

Using s_1^2 and s_2^2 to estimate σ_1^2 and σ_2^2, and substituting the values of the sample statistics yields:

$$(1.0 - 0.9) \pm 1.645\sqrt{\frac{(1.1)^2}{86} + \frac{(1.1)^2}{91}} \Rightarrow .1 \pm .272 \Rightarrow (-.172, .372)$$

c. We estimate (with 90% confidence) that the difference in mean number of promotions awarded to females in large firms and females in small firms is between $-.172$ and $.372$.

d. To decrease the width of the interval, we can decrease the level of confidence or increase the sample size.

3 Simple Linear Regression

3.1 a.

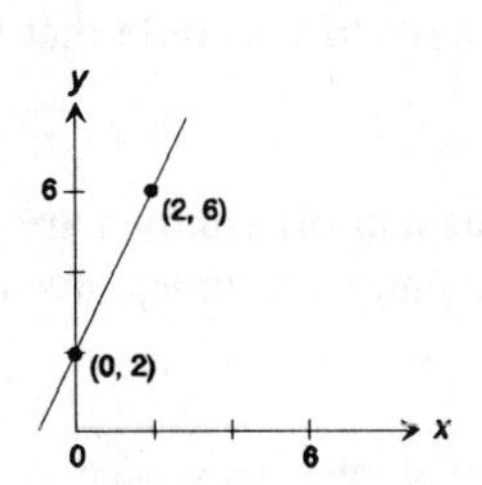

b.

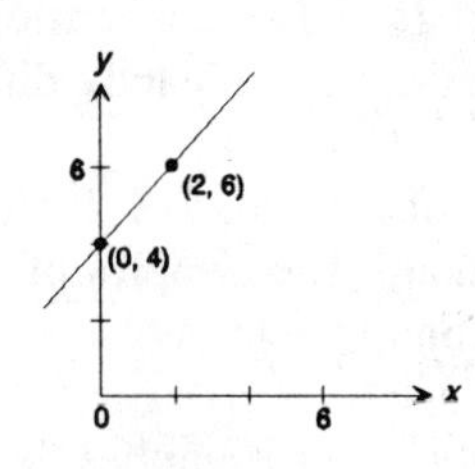

c.

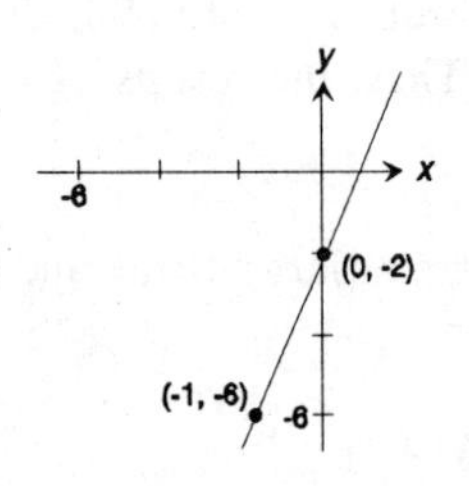

d.

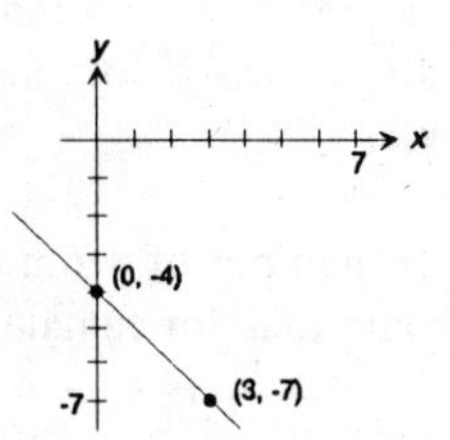

3.3 a. Using the technique explained in Exercise 3.2:

$$\left.\begin{array}{l} 2 = \beta_0 + \beta_1(0) \\ 6 = \beta_0 + \beta_1(2) \end{array}\right\} \rightarrow \left.\begin{array}{l} \beta_0 = 2 \\ \beta_1 = 2 \end{array}\right\} \rightarrow y = 2 + 2x$$

b.

$$\left.\begin{array}{l} 4 = \beta_0 + \beta_1(0) \\ 6 = \beta_0 + \beta_1(2) \end{array}\right\} \rightarrow \left.\begin{array}{l} \beta_0 = 4 \\ \beta_1 = 1 \end{array}\right\} \rightarrow y = 4 + x$$

c.

$$\left.\begin{array}{l} -2 = \beta_0 + \beta_1(0) \\ -6 = \beta_0 + \beta_1(-1) \end{array}\right\} \rightarrow \left.\begin{array}{l} \beta_0 = -2 \\ \beta_1 = 4 \end{array}\right\} \rightarrow y = -2 + 4x$$

d.

$$\left.\begin{array}{l} -4 = \beta_0 + \beta_1(0) \\ -7 = \beta_0 + \beta_1(3) \end{array}\right\} \rightarrow \left.\begin{array}{l} \beta_0 = -4 \\ \beta_1 = -1 \end{array}\right\} \rightarrow y = -4 - x$$

3.5		Slope (β_1)	y-intercept (β_0)
	a.	2	3
	b.	1	1
	c.	3	−2
	d.	5	0
	e.	−2	4

3.7 a. To compute $\hat{\beta}_0$ and $\hat{\beta}_1$, we first construct the following table:

x	y	xy	x^2	y^2
−2	4	−8	4	16
−1	3	−3	1	9
0	3	0	0	9
1	1	1	1	1
2	−1	−2	4	1
$\sum x = 0$	$\sum y = 10$	$\sum xy = -12$	$\sum x^2 = 10$	$\sum y^2 = 36$

Then,

$$SS_{xx} = \sum x^2 - \frac{\left(\sum x\right)^2}{n} = 10 - \frac{(0)^2}{5} = 10$$

$$SS_{xy} = \sum xy - \frac{\left(\sum x\right)\left(\sum y\right)}{n} = -12 - \frac{0(10)}{5} = -12$$

$$\bar{y} = \frac{\sum y}{n} = \frac{10}{5} = 2 \qquad \bar{x} = \frac{\sum x}{n} = \frac{0}{5} = 0$$

Thus, the least squares estimates of β_0 and β_1 are:

$$\hat{\beta}_1 = \frac{SS_{xy}}{SS_{xx}} = \frac{-12}{10} = -1.2$$

$$\hat{\beta}_0 = \bar{y} - \hat{\beta}_1\bar{x} = 2 - (-1.2)(0) = 2$$

and the equation of the least squares prediction line is $\hat{y} = 2 - 1.2x$.

b.

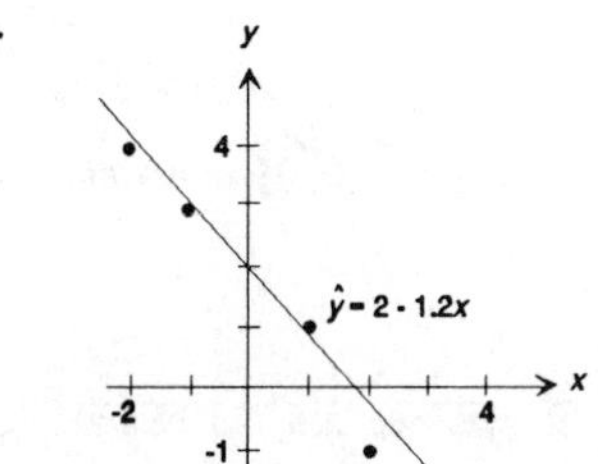

3.9 a. Using Minitab, the scattergram of the data is:

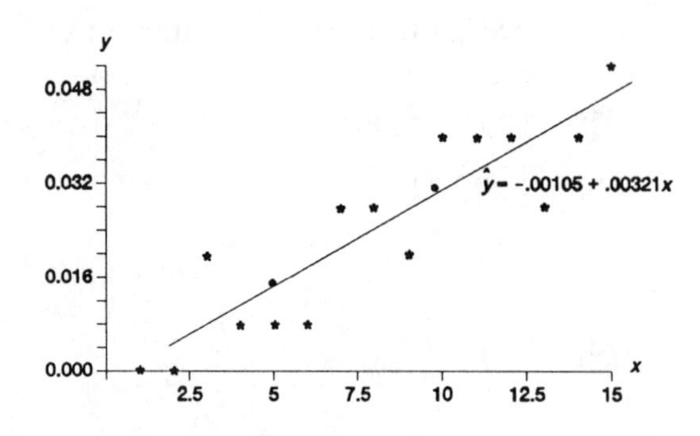

b. Some preliminary calculations:

$\sum x = 120 \quad \sum y = .37 \quad \sum xy = 3.86 \quad \sum x^2 = 1240 \quad \sum y^2 = .0127$

$$SS_{xy} = \sum xy - \frac{\sum x \sum y}{n} = 3.86 - \frac{120(.37)}{15} = .9$$

$$SS_{xx} = \sum x^2 - \frac{(\sum x)^2}{n} = 1240 - \frac{120^2}{15} = 280$$

$$\bar{x} = \frac{\sum x}{n} = \frac{120}{15} = 8 \qquad \bar{y} = \frac{\sum y}{n} = \frac{.37}{15} = .0246667$$

$$\hat{\beta}_1 = \frac{SS_{xy}}{SS_{xx}} = \frac{.9}{280} = .003214285 \approx .00321$$

$$\hat{\beta}_0 = \bar{y} - \hat{\beta}_1\bar{x} = .0246667 - .003214285(8) = -.001047618 \approx -.00105$$

The least squares line is $\hat{y} = -.00105 + .00321x$.

c. See the graph in part (a).

d. $\hat{\beta}_0 = -.00105$. Since $x = 0$ is not in the observed range, $\hat{\beta}_0$ is simply the y-intercept.

$\hat{\beta}_1 = .00321$. When the number of vehicles increases by one, the mean congestion time is estimated to increase by .00321 minutes.

3.11 a. The scattergram of the data is:

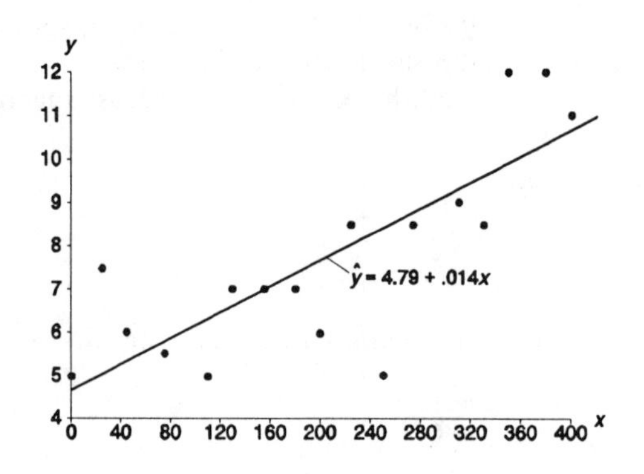

b. Some preliminary calculations are:

$n = 17$

$\sum x = 3395$ $\quad \sum y = 130.27$

$\sum x^2 = 931{,}025$ $\quad \sum y^2 = 1081.3975$ $\quad \sum xy = 29{,}656.15$

Then,

$$SS_{xx} = \sum x^2 - \frac{\left(\sum x\right)^2}{n} = 931{,}025 - \frac{3395^2}{17} = 253{,}023.5294$$

$$SS_{xy} = \sum xy - \frac{\left(\sum x\right)\left(\sum y\right)}{n} = 29{,}656.15 - \frac{3395(130.27)}{17} = 3640.4647$$

$$\bar{x} = \frac{\sum x}{n} = \frac{3395}{17} = 199.7058824 \qquad \bar{y} = \frac{\sum y}{n} = \frac{130.27}{17} = 7.662941176$$

Thus, the least squares estimates are:

$$\hat{\beta}_1 = \frac{SS_{xy}}{SS_{xx}} = \frac{3640.4647}{253{,}023.5294} = .01438785 \approx .014$$

and

$$\hat{\beta}_0 = \bar{y} - \hat{\beta}_1\bar{x} = 7.662941176 - (.01438785)(199.7058824) = 4.7896 \approx 4.79$$

Finally, the least squares line is $\hat{y} = 4.79 + .014x$.

c. See the plot in part (a) for the least squares line.

d. $\hat{\beta}_0 = 4.79$. We estimate the mean time to drill 5 feet when drilling starts at 0 feet to be 4.79 minutes.

$\hat{\beta}_1 = .014$. We estimate the mean time to drill 5 feet will increase .014 minutes for each additional foot of starting depth.

3.13 a. The scattergram is:

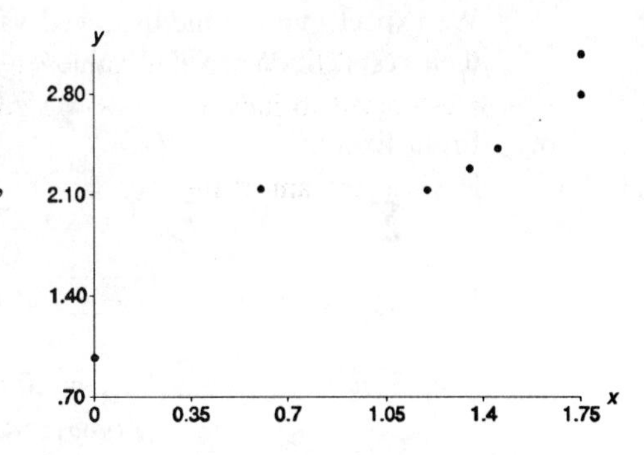

It does appear that the relationship between shear strength and precompression stress is linear. As precompression stress increases, the shear strength increases.

b. A summary of the data enables us to compute the least squares line:

$n = 7$

$\sum x = 8.06$ $\quad \sum y = 16.3$

$\sum x^2 = 11.7388$ $\quad \sum y^2 = 40.6022$ $\quad \sum xy = 21.195$

$$SS_{xx} = \sum x^2 - \frac{\left(\sum x\right)^2}{n} = 11.7388 - \frac{8.06^2}{7} = 2.4582857$$

$$SS_{xy} = \sum xy - \frac{\left(\sum x\right)\left(\sum y\right)}{n} = 21.195 - \frac{8.06(16.3)}{7} = 2.426714$$

$$\bar{x} = \frac{\sum x}{n} = \frac{8.06}{7} = 1.151428571 \qquad \bar{y} = \frac{\sum y}{n} = \frac{16.3}{7} = 2.328571429$$

Hence,

$$\hat{\beta}_1 = \frac{2.426714}{2.4582857} = .987157025 \approx .987$$

$$\hat{\beta}_0 = \bar{y} - \hat{\beta}_0\bar{x} = 2.328571429 - .987157025(1.151428571) = 1.19193 \approx 1.192$$

The least squares line is $\hat{y} = 1.192 + .987x$.

c. $\hat{\beta}_0 = 1.192$. We estimate that the mean shear strength is 1.192 when precompression stress is 0.

$\hat{\beta}_1 = .987$. We estimate that the mean shear strength will increase by .987 for each unit increase in precompression stress.

3.15 a. From Exercise 3.7,

$$\sum y = 10, \ \sum y^2 = 36, \ SS_{xy} = -12, \ \hat{\beta}_1 = -1.2$$

$$SS_{yy} = \sum y^2 - \frac{\left(\sum y\right)^2}{n} = 36 - \frac{10^2}{5} = 16$$

$$SSE = SS_{yy} - \hat{\beta}_1 SS_{xy} = 16 - (-1.2)(-12) = 1.6$$

$$s^2 = \frac{SSE}{n-2} = \frac{1.6}{5-2} = .53333 \qquad s = \sqrt{.53333} = .730$$

We expect most of the observed values of y to fall within $2s$ or 2(.730) or 1.46 units of their respective predicted values.

b. From Exercise 3.9,

$$\sum y = .37, \ \sum y^2 = .0127, \ SS_{xy} = .9, \ \hat{\beta}_1 = .003214285$$

$$SS_{yy} = \sum y^2 - \frac{\left(\sum y\right)^2}{n} = .0127 - \frac{.37^2}{15} = .0035733$$

$$SSE = SS_{yy} - \hat{\beta}_1 SS_{xy} = .0035733 - (.003214285)(.9) = .000680476$$

$$s^2 = \frac{SSE}{n-2} = \frac{.000680476}{15-2} = .000052344 \qquad s = \sqrt{.000052344} = .00723$$

We expect most of the observed values of y to fall within $2s$ or 2(.00723) or .01446 units of their respective predicted values.

c. From Exercise 3.11,

$$\sum y = 130.27, \ \sum y^2 = 1081.3975, \ SS_{xy} = 3640.4647, \ \hat{\beta}_1 = .01438785$$

$$SS_{yy} = \sum y^2 - \frac{(\sum y)^2}{n} = 1081.3975 - \frac{130.27^2}{17} = 83.1461529$$

$$SSE = SS_{yy} - \hat{\beta}_1 SS_{xy} = 83.1461529 - (.01438785)(3640.4647) = 30.76769287$$

$$s^2 = \frac{SSE}{n - 2} = \frac{30.76769287}{17 - 2} = 2.05118 \qquad s = \sqrt{2.05188} = 1.432$$

We expect most of the observed values of y to fall within $2s$ or 2(1.432) or 2.864 units of their respective predicted values.

d. From Exercise 3.13,

$$\sum y = 16.3, \ \sum y^2 = 40.6022, \ SS_{xy} = 2.426714, \ \hat{\beta}_1 = .987157025$$

$$SS_{yy} = \sum y^2 - \frac{(\sum y)^2}{n} = 40.6022 - \frac{16.3^2}{7} = 2.64648571$$

$$SSE = SS_{yy} - \hat{\beta}_1 SS_{xy} = 2.64648571 - (.987157025)(2.426714) = .250937937$$

$$s^2 = \frac{SSE}{n - 2} = \frac{.250937937}{7 - 2} = .05019 \qquad s = \sqrt{.015019} = .2240$$

We expect most of the observed values of y to fall within $2s$ or 2(.224) or .448 units of their respective predicted values.

3.17 a.

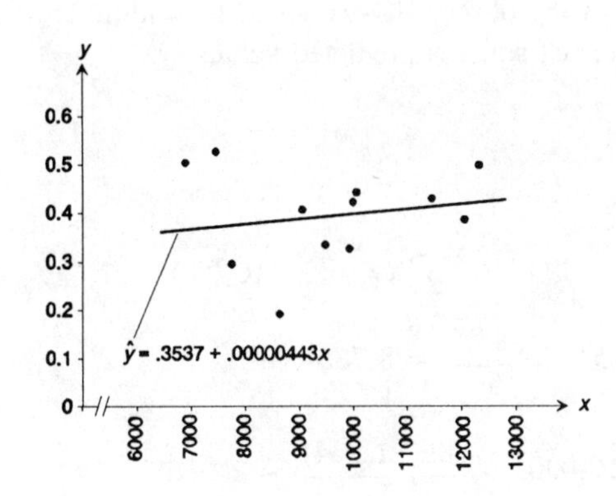

b. We first calculate the following summary information:

$n = 23$

$$\sum x = 115077 \qquad \sum y = 4.754$$

$$\sum x^2 = 1{,}137{,}630{,}209 \qquad \sum y^2 = 1.986222 \qquad \sum xy = 45740.475$$

$$SS_{xx} = \sum x^2 - \frac{(\sum x)^2}{n} = 1{,}137{,}630{,}209 - \frac{(115077)^2}{12} = 34{,}070{,}548$$

$$SS_{xy} = \sum xy - \frac{(\sum x)(\sum y)}{n} = 45740.475 - \frac{(115077)(4.754)}{12} = 150.8035$$

$$SS_{yy} = \sum y^2 - \frac{(\sum y)^2}{n} = 1.986222 - \frac{(4.754)^2}{12} = .10285$$

$$\bar{x} = \frac{\sum x}{n} = \frac{115077}{12} = 9589.75 \qquad \bar{y} = \frac{\sum y}{n} = \frac{4.754}{12} = .39617$$

$$\hat{\beta}_1 = \frac{SS_{xy}}{SS_{xx}} = \frac{150.8035}{34{,}070{,}548} = .000004426$$

$$\hat{\beta}_0 = \bar{y} - \hat{\beta}_1\bar{x} = .39617 - (.000004426)(9589.75) = .3537$$

The equation of the least squares prediction line is $\hat{y} = .3537 + .000004426x$.

c. The line is plotted on the graph of part (a).

d. When $x = \$8{,}000$, we predict the refusal rate will be:

$$\hat{y} = .3537 + .000004426(8000) = .3892$$

e. $SSE = SS_{yy} - \hat{\beta}_1 SS_{xy} = .10285 - (.000004426)(150.8035) = .10218$

$$s^2 = \frac{SSE}{n-2} = \frac{.10218}{12-2} = .01022$$

f. $s = \sqrt{s^2} = .01022 = .10108$

The standard deviation, s, is the estimated value of σ, the standard deviation of the random error ϵ. We expect most of the observed y-values to lie within $2s$ or $2(.10108)$ or $.20216$ units of their respective least squares predicted values, $\hat{y}$.

3.19 a. Some preliminary calculations are:

$$\sum x = 285 \qquad \sum x^2 = 25{,}025$$

$$\sum y = -1{,}964 \qquad \sum y^2 = 771{,}994 \qquad \sum xy = -110{,}030$$

$$SS_{xx} = \sum x^2 - \frac{(\sum x)^2}{n} = 25{,}025 - \frac{(285)^2}{5} = 8{,}780$$

$$SS_{xy} = \sum xy - \frac{\sum x \sum y}{n} = -110{,}030 - \frac{285(-1{,}964)}{5} = 1{,}918$$

$$\bar{x} = \frac{\sum x}{n} = \frac{285}{5} = 57 \qquad \bar{y} = \frac{\sum y}{n} = \frac{-1{,}964}{5} = -392.8$$

$$\hat{\beta}_1 = \frac{SS_{xy}}{SS_{xx}} = \frac{1{,}918}{8{,}780} = .218451025 \approx .2185$$

$$\hat{\beta}_0 = \bar{y} - \hat{\beta}_1\bar{x} = -392.8 - (.218451025)(57) = -405.2517084 \approx -405.25$$

The least squares line is $\hat{y} = -405.25 + .2185x$.

b.

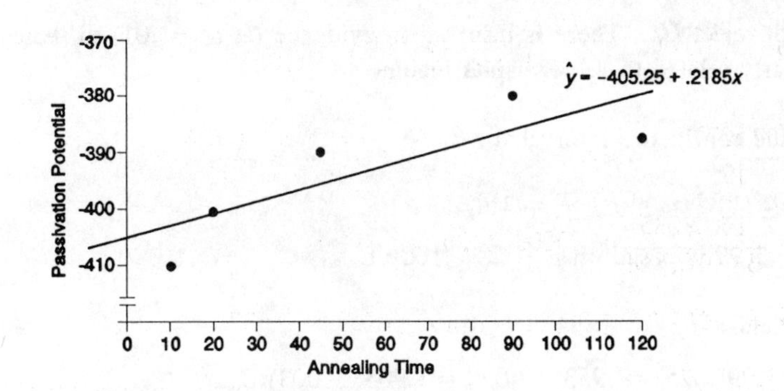

c. $SS_{yy} = \sum y^2 - \dfrac{(\sum y)^2}{n} = 771{,}994 - \dfrac{(-1{,}964)^2}{5} = 534.8$

$SSE = SS_{yy} - \hat{\beta}_1 SS_{xy} = 534.8 - (.218451025)(1{,}918) = 115.8109436$

$s^2 = \dfrac{SSE}{n-2} = \dfrac{115.8109436}{5-2} = 38.6036$

d. $s = \sqrt{38.6036} = 6.213$

We expect most of the observed values of y to fall within $2s$ or $2(6.213)$ or 12.426 units of their respective predicted values.

3.21 From Exercises 3.11 and 3.15, $s = 1.432$, $\hat{\beta}_1 = .0143879$, $SS_{xx} = 253{,}023.5294$.

To determine if dry drill time y increases with depth, we test:

$H_0\colon \beta_1 = 0$
$H_a\colon \beta_1 > 0$

Test statistic: $t = \dfrac{\hat{\beta}_1}{s/\sqrt{SS_{xx}}} = \dfrac{.0143879}{1.432/\sqrt{253{,}023.5294}} = 5.05$

Rejection region: $\alpha = .10$, $df = n - 2 = 15$, $t_{.10} = 1.341$
Reject H_0 if $t > 1.341$

Conclusion: Reject H_0. There is sufficient evidence to indicate that dry drill time y increases with depth x at $\alpha = .10$.

3.23 $H_0\colon \beta_1 = 0$
$H_a\colon \beta_1 \neq 0$

Test statistic: $t = \dfrac{\hat{\beta}_1}{s/\sqrt{SS_{xx}}} = \dfrac{.000004426}{.10108/\sqrt{34{,}070{,}548}} = .256$

Rejection region: $\alpha = .01$, $df = n - 2 = 10$, $t_{.005} = 3.169$
Reject H_0 if $t < -3.169$ or $t > 3.169$

Conclusion: Do not reject H_0. There is insufficient evidence (at $\alpha = .01$) to indicate that y, refusal rate, is linearly related to x, per capita income.

3.25 a. The form of the confidence interval for β_1 is:

$$\hat{\beta}_1 \pm t_{\alpha/2} s/\sqrt{SS_{xx}} \text{ or } \hat{\beta}_1 \pm t_{\alpha/2} s_{\hat{\beta}_1}$$

where $t_{.025} = 1.99$ is based on $n - 2 = 100$ df.

Substitution yields:

$$.953 \pm 1.99(.025) \Rightarrow .953 \pm .050 \Rightarrow (.903, 1.003)$$

b. We are 95% confident β_1 is between .903 and 1.003. This means we are 95% confident the change in the mean theophylline concentration as determined by Ames Seralyzer for each unit change in the theophylline concentration as determined by EMIT is between .903 and 1.003. Since this interval contains 1, it appears that the two methods are very consistent in the measurement of theophylline concentration.

3.27 To determine if the regression models are useful, we test:

H_0: $\beta_1 = 0$
H_a: $\beta_1 \neq 0$

Rejection region: $\alpha = .05$, df $= n - 2 = 18$, $t_{.025} = 2.101$
Reject H_0 if $t < -2.101$ or $t > 2.101$

a. Since the observed value of the test statistic falls in the rejection region ($t = 7.13 > 2.101$), H_0 is rejected. There is sufficient evidence to indicate the two models for cogener 2, 3, 4, 7, 8-P_n CDF are useful at $\alpha = .05$.

b. Since the observed value of the test statistic falls in the rejection region ($t = 5.98 > 2.101$), H_0 is rejected. There is sufficient evidence to indicate the two models for cogener H_x CDD are useful at $\alpha = .05$.

c. Since the observed value of the test statistic falls in the rejection region ($t = 4.98 > 2.101$), H_0 is rejected. There is sufficient evidence to indicate the two models for cogener OCDD are useful at $\alpha = .05$.

d. For a fat level of 8.0 ppm, the estimate of the level of 2, 3, 4, 7, 8-P_n CDF in blood plasma is:

$$\hat{y} = .9855 + .7605(8) = 7.0695 \text{ ppm}$$

e. For a blood plasma level of 24.0 ppm, the estimate of the level of H_xCDD in fat tissue is:

$$\hat{y} = 18.1565 + .7377(24.0) = 35.8613 \text{ ppm}$$

f. For a fat level of 776 ppm, the estimate of the level of OCDD in blood plasma is:

$$\hat{y} = 167.723 + 1.5752(776) = 1390.08 \text{ ppm}$$

3.29 a.

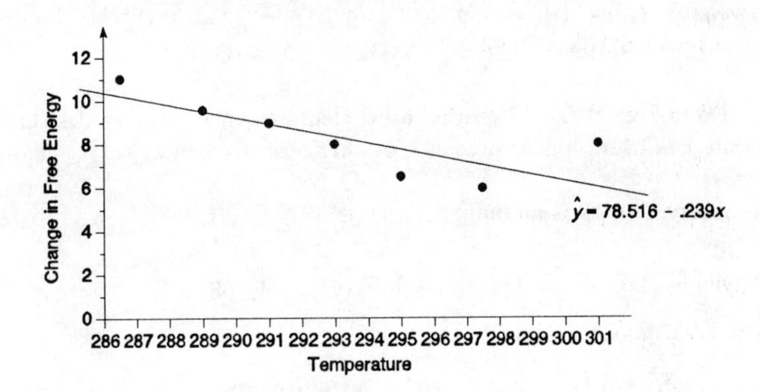

b. Some preliminary calculations are:

$\sum x = 2{,}053 \qquad \sum x^2 = 602{,}265.5$

$\sum y = 59.2 \qquad \sum y^2 = 517.22 \qquad \sum xy = 17{,}326.7$

$$SS_{xx} = \sum x^2 - \frac{\left(\sum x\right)^2}{n} = 602{,}265.5 - \frac{(2{,}053)^2}{7} = 149.92857$$

$$SS_{xy} = \sum xy - \frac{\sum x \sum y}{n} = 17{,}326.7 - \frac{(2{,}053)(59.2)}{7} = -35.814286$$

$$SS_{yy} = \sum y^2 - \frac{\left(\sum y\right)^2}{n} = 517.22 - \frac{59.2^2}{7} = 16.5571429$$

$$\bar{x} = \frac{\sum x}{n} = \frac{2{,}053}{7} = 293.28571 \qquad \bar{y} = \frac{\sum y}{n} = \frac{59.2}{7} = 8.4571429$$

$$\hat{\beta}_1 = \frac{SS_{xy}}{SS_{xx}} = \frac{-35.814286}{149.92857} = -.2388757 \approx -.239$$

$$\hat{\beta}_0 = \bar{y} - \hat{\beta}_1 \bar{x} = 78.516$$

The least squares line is $\hat{y} = 78.516 - .239x$.

c. The least squares line is plotted on the graph above.

d. $SSE = SS_{yy} = \hat{\beta}_1 SS_{xy} = 16.5571429 - (-.2388757)(-35.814286) = 8.001980262$

$$s^2 = \frac{SSE}{n-2} = \frac{8.001980262}{7-2} = 1.600396 \qquad s = \sqrt{1.600396} = 1.265$$

To determine if temperature is a useful linear predictor of change in free energy, we test:

H_0: $\beta_1 = 0$
H_a: $\beta_1 \neq 0$

Test statistic: $t = \dfrac{\hat{\beta}_1 - 0}{s/\sqrt{SS_{xx}}} = \dfrac{-.239 - 0}{1.265/\sqrt{149.9286}} = -2.31$

Rejection region: $\alpha = .01$, $n - 2 = 7 - 2 = 5$, $t_{.005} = 4.032$
Reject H_0 if $t < -4.032$ or $t > 4.032$

Conclusion: Do not reject H_0. There is insufficient evidence (at $\alpha = .01$) to indicate that temperature is a useful linear predictor of change in free energy.

e. It looks like observation #5 is an outlier.

f. $\sum x = 1{,}752$ $\quad \sum x^2 = 511{,}664.5$ $\quad \sum xy = 14{,}768.2$

$\sum y = 50.7$ $\quad \sum y^2 = 444.97$

$$SS_{xy} = \sum xy - \frac{(\sum x)(\sum y)}{n} = 14{,}768.2 - \frac{(1{,}752)(50.7)}{6} = -36.2$$

$$SS_{xx} = \sum x^2 - \frac{(\sum x)^2}{n} = 511{,}664.5 - \frac{(1{,}752)^2}{6} = 80.5$$

$$\bar{x} = \frac{\sum x}{n} = \frac{1{,}752}{6} = 292 \qquad \bar{y} = \frac{\sum y}{n} = \frac{50.7}{6} = 8.45$$

$$\hat{\beta}_1 = \frac{SS_{xy}}{SS_{xx}} = \frac{-36.2}{80.5} = -.449689441 \approx -.4497$$

$$\hat{\beta}_0 = \bar{y} - \hat{\beta}_1\bar{x} = 8.45 - (-.449689441)(292) = 139.7593168 \approx 139.759$$

The new least squares line is $\hat{y} = 139.759 - .4497x$.

$$SS_{yy} = \sum y^2 - \frac{(\sum y)^2}{n} = 444.97 - \frac{(50.7)^2}{6} = 16.555$$

$$SSE = SS_{yy} - \hat{\beta}_1 SS_{xy} = 16.555 - (-.449689441)(-36.2) = .276242235$$

$$s^2 = \frac{SSE}{n-2} = \frac{.276242235}{6-2} = .06906 \qquad s = \sqrt{.06906} = .2628$$

To determine if temperature is a useful predictor of change in free energy, we test:

H_0: $\beta_1 = 0$
H_a: $\beta_1 \neq 0$

Test statistic: $$t = \frac{\hat{\beta}_1 - 0}{s/\sqrt{SS_{xx}}} = \frac{-.4497 - 0}{.2628/\sqrt{80.5}} = -15.35$$

Rejection region: $\alpha = .01$, df $= n - 2 = 4$, $t_{.005} = 4.604$
Reject H_0 if $t < -4.604$ or $t > 4.604$

Conclusion: Reject H_0. There is sufficient evidence to indicate temperature is a useful predictor of change in free energy at $\alpha = .01$.

3.31 We would expect the GPA of a college student to be correlated to his/her I.Q. As the I.Q. score increases, we would expect the GPA to increase. Thus, the correlation would be positive.

3.33 a. An example of positively correlated variables is price of a product and the quantity supplied.

b. An example of negatively correlated variables is price of a product and the quantity demanded.

3.35 a. $\hat{\beta}_0 = 175.4$: We estimate the mean time required to perform the task to equal 175.4 milliseconds when the Index of Difficulty is 0.

$\hat{\beta}_1 = 133.2$: For every one unit increase in the Index of Difficulty, we estimate the mean time required to perform the task to increase 133.2 milliseconds.

b. $r = .951$: There is a strong positive linear relationship between the Index of Difficulty and the time required to perform the task.

c. To determine if the Index of Difficulty is a useful linear predictor of time, we test:

H_0: $\rho = 0$
H_a: $\rho \neq 0$

Test statistic: $t = \dfrac{r\sqrt{n-2}}{\sqrt{1-r^2}} = \dfrac{.951\sqrt{160-2}}{\sqrt{1-.951^2}} = 38.662$

Rejection region: $\alpha = .05$, df $= n - 2 = 158$, $t_{.025} \approx 1.96$
Reject H_0 if $t < -1.96$ or $t > 1.96$

Conclusion: Reject H_0. There is sufficient evidence (at $\alpha = .05$) to indicate the Fitt's Law Model is statistically adequate for predicting performance time.

d. $r^2 = (.951)^2 = .9044$

90.44% of the sum of squares of deviation of the sample time values about their mean can be explained by using the Index of Difficulty as a linear predictor.

3.37 a. $r^2 = .587$. 58.7% of the variability in the state's chemical industry E/J ratio can be explained by the linear relationship between the state's chemical industry E/J ratio and the state's pollution abatement capital expenditures.

b. $r = -\sqrt{.587} = -.766$. (The value of r is negative since the slope is negative.) There is a mildly strong negative linear relationship between the state's chemical industry E/J ratio and the state's pollution abatement capital expenditures.

c. H_0: $\rho = 0$
H_a: $\rho < 0$

Test statistic: $t = \dfrac{r\sqrt{n-2}}{\sqrt{1-r^2}} = \dfrac{-.766\sqrt{19-2}}{\sqrt{1-.587}} = -4.91$

Rejection region: $\alpha = .01$, df $= n - 2 = 17$, $t_{.01} = 2.567$
Reject H_0 if $t < -2.567$

Conclusion: Reject H_0. There is sufficient evidence to indicate (at $\alpha = .01$) that underspending on pollution control will result in higher emissions and fewer jobs.

3.39 We wish to test:

H_0: $\rho = 0$
H_a: $\rho > 0$

Test statistic: $t = \dfrac{r\sqrt{n-2}}{\sqrt{1-r^2}} = \dfrac{.46\sqrt{58-2}}{\sqrt{1-.46^2}} = 3.88$

Rejection region: $\alpha = .01$, df $= n - 2 = 56$, $t_{.01} \approx 2.390$
Reject H_0 if $t > 2.390$

Conclusion: Reject H_0. There is sufficient evidence (at $\alpha = .01$) to indicate there is a positive correlation between group maturity and outside-of-class meetings.

3.41 a. $\hat{\beta}_1 = 450$. For each additional processing step, the mean response time is estimated to increase by 450 milliseconds.

b. $r^2 = .91$. This means that 91% of the total sum of squares of deviations of the response times about their means can be attributed to the linear relationship between response time and number of processing steps.

c. $r = \sqrt{.91} = .9539$. To determine if the model is adequate, we test:

H_0: $\rho = 0$
H_a: $\rho \neq 0$

Test statistic: $t = \dfrac{r\sqrt{n-2}}{\sqrt{1-r^2}} = \dfrac{.9539\sqrt{8-2}}{\sqrt{1-.91}} = 7.79$

Rejection region: $\alpha = .01$, df $= n - 2 = 6$, $t_{.005} = 3.707$
Reject H_0 if $t < -3.707$ or $t > 3.707$

Conclusion: Reject H_0. There is sufficient evidence to indicate the model is adequate at $\alpha = .05$.

3.43 a. From Exercises 3.9 and 3.15, $\hat{y} = -.00105 + .00321x$, $s = .00723$, $SS_{xx} = 280$, $\bar{x} = 8$. If there are 10 vehicles, then $x_p = 10$. When $x_p = 10$, $\hat{y} = -.00105 + .00321(10) = .03105$.

A 90% prediction interval for the congestion time is $\hat{y} \pm t_{.05}s\sqrt{1 + \dfrac{1}{n} + \dfrac{(x_p - \bar{x})^2}{SS_{xx}}}$

where $\hat{y} = .03105$, $t_{.05} = 1.771$ is based on $n - 2 = 13$, $s = .00723$, $n = 15$, $\bar{x} = 8$, and $SS_{xx} = 280$.

Substituting, we have:

$$.03105 \pm 1.771(.00723)\sqrt{1 + \frac{1}{15} + \frac{(10 - 8)^2}{280}} \Rightarrow .03105 \pm .01331$$
$$\Rightarrow (.01774, .04436)$$

We are 90% confident that the actual congestion time when 10 vehicles are operated will fall between .01774 and .04436.

b. A 90% confidence interval for the mean congestion time is $\hat{y} \pm t_{.05}s\sqrt{\frac{1}{n} + \frac{(x_p - \bar{x})^2}{SS_{xx}}}$

Substituting, we have:

$$.03105 \pm 1.771(.00723)\sqrt{\frac{1}{15} + \frac{(10 - 8)^2}{280}} \Rightarrow .03105 \pm .00364 \Rightarrow (.02741, .03469)$$

We are 90% confident that the mean congestion time when 10 vehicles are operated will fall between .02741 and .03469.

c. If n is increased, the critical value of t will decrease, thus causing the interval to be smaller. Also, since $1/n$ appears in both the formula for the prediction interval and the confidence interval, by increasing n, the value of $1/n$ decreases. This, too, will cause the intervals to be smaller. (This assumes that the value of s is the same.)

3.45 a. From Exercise 3.12, $\hat{y} = 4.4417 + .2612x$.

$$\sum y^2 = 713.5745 \quad \sum y = 101.89 \quad \bar{x} = 9$$
$$SS_{xy} = 73.14 \quad SS_{xx} = 280$$
$$SS_{yy} = \sum y^2 - \frac{(\sum y)^2}{n} = 713.5745 - \frac{101.89^2}{15} = 21.4696933$$
$$SSE = SS_{yy} - \hat{\beta}_1 SS_{xy} = 21.4696933 - .2612143(73.14) = 2.36448$$
$$s^2 = \frac{SSE}{n - 2} = \frac{2.36448}{15 - 2} = .181883 \qquad s = \sqrt{s^2} = \sqrt{.181883} = .4265$$

The form of the confidence interval for $E(y)$ is $\hat{y} \pm t_{\alpha/2}s\sqrt{\frac{1}{n} + \frac{(x - \bar{x})^2}{SS_{xx}}}$

For $x = 7$, $\hat{y} = 4.4417 + .2612(7) = 6.2701$

For confidence coefficient .95, $\alpha = 1 - .95 = .05$ and $\alpha/2 = .05/2 = .025$. From Table 2, Appendix C, $t_{.025} = 2.160$ with df $= n - 2 = 15 - 2 - 13$.

The confidence interval is:

$$6.2701 \pm 2.160(.4265)\sqrt{\frac{1}{15} + \frac{(7 - 9)^2}{280}} \Rightarrow 6.2701 \pm .2621 \Rightarrow (6.008, 6.5322)$$

We are 95% confident that the mean average peak EEG frequency is between 6.008 and 6.5322 when age = 7.

b. The form of the prediction interval is:

$$\hat{y} \pm t_{\alpha/2} s \sqrt{1 + \frac{1}{n} + \frac{(x - \bar{x})^2}{SS_{xx}}} \Rightarrow 6.2701 \pm 2.160(.4265)\sqrt{1 + \frac{1}{15} + \frac{(7 - 9)^2}{280}}$$

$$\Rightarrow 6.2701 \pm .9578 \Rightarrow (5.3123, 7.2279)$$

We are 95% confident that the actual average peak EEG frequency is between 5.3123 and 7.2279 when age = 7.

3.47 A 95% prediction interval for the refusal rate, y, is $\hat{y} \pm t_{.025} s \sqrt{1 + \frac{1}{n} + \frac{(x_p - \bar{x})^2}{SS_{xx}}}$

where $t_{.025} = 2.228$ is based on df $= n - 2 = 10$.

From Exercise 3.17, $\hat{y} = .3537 + .000004426x$, $\bar{x} = 9589.75$, $s = .1011$, $n = 12$, and $SS_{xx} = 34{,}070{,}548$. For $x_p = 8{,}000$, $\hat{y} = .3537 + .000004426(8{,}000) = .3891$.

Substituting, we have:

$$.3891 \pm 2.228(.1011)\sqrt{1 + \frac{1}{12} + \frac{(8000 - 9589.75)^2}{34{,}070{,}548}} \Rightarrow .3891 \pm .2423 \Rightarrow (.1468, .6314)$$

We predict the actual refusal rate will fall between .1468 and .6314 when per capital income is \$8,000 with 95% confidence.

This interval is so wide because the model is not adequate for predicting refusal rate.

3.49 From Exercise 3.19, $\hat{y} = -405.25 + .2135x$, $s = 6.213$, $SS_{xx} = 8780$, $n = 5$, and $\bar{x} = 57$.

The 95% confidence interval for the mean passivation potential of crystallized alloy when annealing time is 30 minutes is:

$$\hat{y} \pm t_{\alpha/2} s \sqrt{\frac{1}{n} + \frac{(x_p - \bar{x})^2}{SS_{xx}}} \quad \text{where } t_{.025} = 3.182 \text{ with df} = n - 2 = 3.$$

$$\hat{y} = \hat{\beta}_0 + \hat{\beta}_1 x = -405.25 + .2185(30) = -398.70$$

Using substitution:

$$-398.70 \pm 3.182(6.213)\sqrt{\frac{1}{5} + \frac{(30 - 57)^2}{8780}} \Rightarrow -398.70 \pm 10.513$$

$$\Rightarrow (-409.218, -388.182)$$

We are 95% confident that when the annealing time is 30 minutes, the mean passivation potential of crystallized alloy is between -409.218 and -388.182.

3.51 From the printout, we find the following:

Least squares prediction line: $\hat{y} = 28.894 + .095x$

SSE = 13,236.45610; s^2 = MSE = 315.15372; s = Root MSE = 17.75257

Most of the observed values of the CPU times will fall within $2s$ or 2(17.75257) or 35.50514 seconds of their respective predicted values.

r^2 = R-SQUARE = .0661. This indicates that only 6.61% of the variability in the CPU times can be explained by the linear relationship between the CPU time and the I/O units.

The test statistic for testing if the model is adequate for predicting CPU time (H_0: $\beta_1 = 0$) is $t = 1.725$ and the p-value is .0919. If $\alpha = .05$, we would not reject H_0. There is insufficient evidence to indicate the model is adequate for predicting CPU time.

3.53 From the printout, we find the following:

Least squares prediction line: $\hat{y} = -.5349 + 15.5262x$

SSE = 6.97415579; s^2 = MSE = .87176947; s = Root MSE = .933686

Most of the observed values of the carbon monoxide contents will fall within $2s$ or 2(.933686) or 1.867372 units of their respective predicted values.

r^2 = R-SQUARE = .9760. This indicates that 97.6% of the variability in the carbon monoxide contents can be explained by the linear relationship between the carbon monoxide contents and the nicotine contents.

The test statistic for testing if the model is adequate for predicting carbon monoxide contents (H_0: $\beta_1 = 0$) is $t = 18.047$ and the p-value is .0001. If $\alpha \geq .0001$, we would reject H_0. There is sufficient evidence to indicate the model is adequate for predicting carbon monoxide contents.

The 95% prediction interval for carbon monoxide content when the nicotine content is .4 (5th observation) is (3.3570, 7.9942). We are 95% confident that the actual carbon monoxide content will be between 3.3570 and 7.9942 when the nicotine content is .4.

3.55 a. $\hat{y} = .2134 + 2.4264x$

b. r^2 = R SQ. = .837041

c. $t = 10.6303$ (22 df)

d. p-value = 3.92489E-10 $\approx$ .0000. For any value of $\alpha > p$-value, we will reject H_0. Since the p-value is almost 0, we will reject H_0 for any reasonable value of α. There is sufficient evidence to indicate the model for predicting heat transfer enhancement is adequate.

e. s = SD. ER. EST. = .453826. Almost all of the observed values of heat transfer enhancement will be within $2s$ or 2(.453826) or .908 units of their respective predicted values.

3.57 a. The results of the preliminary calculations are provided below:

$$n = 5, \sum x^2 = 30, \sum xy = -278, \sum y^2 = 2589$$

Substituting into the formula for $\hat{\beta}_1$, we have $\hat{\beta}_1 = \dfrac{\sum xy}{\sum x^2} = \dfrac{-278}{30} = -9.2667$, and the least squares line is $\hat{y} = -9.2667x$.

b. $\text{SSE} = \sum y^2 - \hat{\beta}_1 \sum xy = 2589 - (-9.2667)(-278) = 12.8667$

$$s^2 = \frac{\text{SSE}}{n-1} = \frac{12.8667}{5-1} = 3.2167 \qquad s = \sqrt{s^2} = \sqrt{3.2167} = 1.7935$$

c. $H_0\text{: } \beta_1 = 0$
$H_a\text{: } \beta_1 < 0$

$$\text{Test statistic} = t = \frac{\hat{\beta}_1}{s/\sqrt{\sum x^2}} = \frac{-9.2667}{1.7935/\sqrt{30}} = -28.300$$

Rejection region: $\alpha = .05$, $\text{df} = n - 1 = 4$, $t_{.05} = 2.132$
Reject H_0 if $t < -2.132$

Conclusion: Reject H_0. There is sufficient evidence (at $\alpha = .05$) to indicate that x and y are negatively linearly related.

d. A 95% confidence interval for β_1 is $\hat{\beta}_1 \pm t_{.025} s/\sqrt{\sum x^2}$

where $t_{.025} = 2.776$ is based on $n - 1 = 4$ df. Substituting, we have:

$$-9.2667 \pm 2.776 \frac{1.7935}{\sqrt{30}} \Rightarrow -9.2667 \pm .9090 \Rightarrow (-10.1757, -8.3577)$$

e. When $x_p = 1$, $\hat{y} = \hat{\beta}_1 x = (-9.2667)(1) = -9.2667$.

A 95% confidence interval for $E(y)$ is $\hat{y} \pm t_{.025} s \left(\frac{x_p}{\sqrt{\sum x^2}} \right)$

$$\Rightarrow -9.2667 \pm 2.776(1.7935) \left(\frac{1}{\sqrt{30}} \right)$$

$$\Rightarrow -9.2667 \pm .9090 \Rightarrow (-10.1757, -8.3577)$$

f. From part (e), we predict $\hat{y} = -9.2667$ when $x_p = 1$.

A 95% prediction interval for y is $\hat{y} \pm t_{.025} s \sqrt{1 + \left(\frac{x_p^2}{\sum x^2} \right)}$

$$\Rightarrow -9.2667 \pm 2.776(1.7935) \sqrt{1 + \frac{1^2}{30}}$$

$$\Rightarrow -9.2667 \pm 5.0611 \Rightarrow (-14.3278, -4.2056)$$

3.59 a. Preliminary calculations yield:

$$n = 8, \sum x^2 = 59.75, \sum xy = 320.5, \sum y^2 = 1738$$

Then, $\hat{\beta}_1 = \dfrac{\sum xy}{\sum x^2} = \dfrac{320.5}{59.75} = 5.364$, and the least squares line is $\hat{y} = 5.364x$.

b. H_0: $\beta_1 = 0$
H_a: $\beta_1 \neq 0$

Test statistic: $t = \dfrac{\hat{\beta}_1}{s/\sqrt{\sum x^2}}$

where $s = \sqrt{s^2} = \sqrt{\dfrac{SSE}{n-1}} = \sqrt{\dfrac{\sum y^2 - \hat{\beta}_1 \sum xy}{n-1}} = \sqrt{\dfrac{1738 - (5.364)(320.5)}{7}}$
$= 1.640$

Substituting, we have $t = \dfrac{5.364}{1.640/\sqrt{59.75}} = 25.28$

Rejection region: $\alpha = .10$, df $= n - 1 = 7$, $t_{\alpha/2} = t_{.05} = 1.895$
Reject H_0 if $t > 1.895$ or $t < -1.895$

Conclusion: Reject H_0 at $\alpha = .10$. There is sufficient evidence to indicate that the straight-line model through the origin is useful for predicting decrease in pulse rate y.

c. We want to predict the decrease in pulse rate y corresponding to a drug dosage of $x_p = 3.5$ cubic centimeters. First, we obtain the point estimate:
$\hat{y} = \hat{\beta}_1 x = 5.364(3.5) = 18.774$

A 99% prediction interval for y is $\hat{y} \pm t_{.005} s \sqrt{1 + \dfrac{x_p^2}{\sum x^2}}$ where $t_{.005} = 3.499$ is based on $n - 1 = 7$ df.

Substitution yields:

$$18.774 \pm 3.499(1.640)\sqrt{1 + \frac{(3.5)^2}{59.75}} \Rightarrow 18.774 \pm 6.30 \text{ or } (12.474, 25.074)$$

Therefore, we predict the decrease in pulse rate corresponding to a dosage of 3.5 c.c. to fall between 12.474 and 25.074 beats/minute with 99% confidence.

3.61 a. Preliminary calculations yield:

$$n = 10, \sum x^2 = 1{,}933{,}154, \sum xy = 98{,}946{,}257, \sum y^2 = 5{,}066{,}358{,}119$$

Then, $\hat{\beta}_1 = \dfrac{\sum xy}{\sum x^2} = \dfrac{98{,}946{,}257}{1{,}933{,}154} = 51.18$, and the least squares prediction equation is $\hat{y} = 51.18x$.

b. H_0: $\beta_1 = 0$
H_a: $\beta_1 \neq 0$

Test statistic: $t = \dfrac{\hat{\beta}_1}{s/\sqrt{\sum x^2}}$

where $s = \sqrt{s^2} = \sqrt{\dfrac{SSE}{n - 1}} = \sqrt{\dfrac{\left(\sum y^2 - \hat{\beta}_1 \sum xy\right)}{n - 1}}$

$= \sqrt{\dfrac{5{,}066{,}358{,}119 - 51.18385(98{,}946{,}257)}{10 - 1}} = 460.4036$

Substituting, we have $t = \dfrac{51.18}{460.4036/\sqrt{1{,}933{,}154}} = 154.56$

Rejection region: $\alpha = .01$, df $= n - 1 = 9$, $t_{.005} = 3.250$
Reject H_0 if $t < -3.250$ or $t > 3.250$

Conclusion: Reject H_0. There is sufficient evidence (at $\alpha = .01$) to indicate x, population in service area, contributes information for the prediction of y, residential electric customers in service area.

c. We need the following additional information:

$\sum x = 4286$, $\sum y = 220{,}297$
$SS_{xx} = 96174.4$, $SS_{xy} = 4{,}526{,}962.8$, $SS_{yy} = 213281298$
$\hat{\beta}_1 = 47.07$, $\hat{\beta}_0 = 1855.35$, SSE $= 195{,}568.4$
$s^2 = 24{,}446.05$, $s = 156.3523$

The least squares prediction equation is $\hat{y} = 1855.35 + 47.07x$.

H_0: $\beta_1 = 0$
H_a: $\beta_1 \neq 0$

Test statistic: $t = \dfrac{\hat{\beta}_1}{s/\sqrt{SS_{xx}}} = \dfrac{47.07}{156.3523/\sqrt{96174.4}} = 93.36$

Rejection region: $\alpha = .01$, df $= n - 2 = 8$, $t_{.005} = 3.355$
Reject H_0 if $t < -3.355$ or $t > 3.355$

Conclusion: Reject H_0. There is sufficient evidence (at $\alpha = .01$) to indicate that x contributes information for the prediction of y.

d. Without running a formal test, we can compare the two models. The value of s for the model $y = \beta_1 x + \epsilon$ is 460.4036 while the value of s for the model $y = \beta_0 + \beta_1 x + \epsilon$ is 156.3523. Since the value of s is much smaller for the second model, it appears that the second model should be used.

For a formal test, refer to part (d) of Exercise 3.60.

H_0: $\beta_0 = 0$
H_a: $\beta_0 \neq 0$

Test statistic: $t = \dfrac{\hat{\beta}_0 - 0}{s\sqrt{\dfrac{1}{n} + \dfrac{\bar{x}^2}{SS_{xx}}}} = \dfrac{1855.35}{156.3523\sqrt{\dfrac{1}{10} + \dfrac{428.6^2}{96174.4}}} = 8.37$

Rejection region: $\alpha = .01$, df $= n - 2 = 8$, $t_{.005} = 3.355$
Reject H_0 if $t < -3.355$ or $t > 3.355$

Conclusion: Reject H_0. There is sufficient evidence (at $\alpha = .01$) to indicate that β_0 should be included in the model.

3.63 For Scheduling:

To determine if there is a positive linear relationship between scheduling and performance, we test:

H_0: $\rho = 0$
H_a: $\rho > 0$

Test statistic: $t = \dfrac{r\sqrt{n - 2}}{\sqrt{1 - r^2}} = \dfrac{.22\sqrt{122 - 2}}{\sqrt{1 - .22^2}} = 2.47$

Rejection region: $\alpha = .05$, df $= n - 2 = 120$, $t_{.05} = 1.685$
Reject H_0 if $t > 1.685$

Conclusion: Reject H_0. There is sufficient evidence to indicate a positive linear relationship between scheduling and performance at $\alpha = .05$.

For Synchronization:

To determine if there is a positive linear relationship between synchronization and performance, we test:

H_0: $\rho = 0$
H_a: $\rho > 0$

Test statistic: $t = \dfrac{r\sqrt{n - 2}}{\sqrt{1 - r^2}} = \dfrac{.07\sqrt{122 - 2}}{\sqrt{1 - .07^2}} = .77$

From above, the rejection region is $t > 1.685$.

Conclusion: Do not reject H_0. There is insufficient evidence to indicate a positive linear relationship between synchronization and performance at $\alpha = .05$.

For Allocation of time:

To determine if there is a positive linear relationship between allocation of time and performance, we test:

H_0: $\rho = 0$
H_a: $\rho > 0$

Test statistic: $t = \dfrac{r\sqrt{n-2}}{\sqrt{1-r^2}} = \dfrac{-.25\sqrt{122-2}}{\sqrt{1-(-.25)^2}} = -2.83$

From above, the rejection region is $t > 1.685$.

Conclusion: Do not reject H_0. There is insufficient evidence to indicate a positive linear relationship between allocation of time and performance at $\alpha = .05$.

For Autonomy of time use:

To determine if there is a positive linear relationship between autonomy of time use and performance, we test:

H_0: $\rho = 0$
H_a: $\rho > 0$

The test statistic is $t = \dfrac{r\sqrt{n-2}}{\sqrt{1-r^2}} = \dfrac{-.41\sqrt{122-2}}{\sqrt{1-(-.41)^2}} = -4.92$

From above, the rejection region is $t > 1.685$.

Conclusion: Do not reject H_0. There is insufficient evidence to indicate a positive linear relationship between autonomy of time use and performance at $\alpha = .05$.

For Future orientation:

To determine if there is a positive linear relationship between future orientation and performance, we test:

H_0: $\rho = 0$
H_a: $\rho > 0$

The test statistic is $t = \dfrac{r\sqrt{n-2}}{\sqrt{1-r^2}} = \dfrac{.44\sqrt{122-2}}{\sqrt{1-(.44)^2}} = 5.37$

From above, the rejection region is $t > 1.685$.

Conclusion: Reject H_0. There is sufficient evidence to indicate a positive linear relationship between future orientation and performance at $\alpha = .05$.

3.65 a. H_0: $\beta_1 = 0$
H_a: $\beta_1 \neq 0$

where β_1 is the slope of the line relating estimated ME content to actual ME content for cats fed canned foods.

Test statistic: $t = \dfrac{\hat{\beta}_1}{s_{\hat{\beta}_1}} = \dfrac{.96}{.04} = 24.0$

Rejection region: $\alpha = .05$, df $= n - 2 = 26$, $t_{.025} = 2.056$
Reject H_0 if $t < -2.056$ or $t > 2.056$

Conclusion: Reject H_0. There is sufficient evidence (at $\alpha = .05$) to indicate that x, estimated ME content, is a useful predictor of y, actual ME content, for cats fed canned foods.

b. H_0: $\beta_1 = 0$
H_a: $\beta_1 \neq 0$

where β_1 is the slope of the line relating estimated ME content to actual ME content for cats fed dry foods.

Test statistic: $t = \dfrac{\hat{\beta}_1}{s_{\hat{\beta}_1}} = \dfrac{.84}{.09} = 9.333$

Rejection region: $\alpha = .05$, $\text{df} = n - 2 = 27$, $t_{.025} = 2.052$
Reject H_0 if $t < -2.052$ or $t > 2.052$

Conclusion: Reject H_0. There is sufficient evidence (at $\alpha = .05$) to indicate that x, estimated ME content, is a useful predictor of y, actual ME content, for cats fed dry foods.

c. For cats fed canned foods, the coefficient of determination is:

$$r^2 = (.97)^2 = .941$$

The independent variable x, estimated ME content, explains 94.1% of the variability in the dependent variable y, actual ME content, for cats fed canned foods when we use the least squares equation to predict y.

For cats fed dry foods, the coefficient of determination is:

$$r^2 = (.88)^2 = .774.$$

The independent variable x, estimated ME content, explains 77.4% of the variability in the dependent variable y, actual ME content, for cats fed dry foods when we use the least squares equation to predict y.

d. The standard deviation, s, is the estimated value of σ, the standard deviation of the random error, ϵ. We expect most of the observed y-values to lie within $2s$ of their respective least squares predicted values, $\hat{y}$. Since the model for cats fed canned foods has a smaller standard deviation, the observed y-values are closer to the $\hat{y}$-values than in the model for cats fed dry foods.

3.67 a. H_0: $\rho = 0$
H_a: $\rho < 0$

Test statistic: $t = \dfrac{r\sqrt{n - 2}}{\sqrt{1 - r^2}} = \dfrac{-.20\sqrt{1076 - 2}}{\sqrt{1 - (-.20)^2}} = -6.69$

Rejection region: $\alpha = .01$, $\text{df} = n - 2 = 1074$, $t_{.01} = 2.326$
Reject H_0 if $t < -2.326$

Conclusion: Reject H_0. There is sufficient evidence (at $\alpha = .01$) to indicate a negative linear relationship between age of marketers and Machiavellian score.

b. The coefficient of determination is:

$$r^2 = (-.20)^2 = .04$$

The independent variable x, age of marketer, explains only 4% of the variability in the dependent variable y, Machiavellian score, when we use the least squares equation to predict y.

3.69 a.

$n = 15$	$SS_{xx} = 8.0818933$	$\hat{\beta}_1 = -1782.833725$
$\sum x = 46.69$	$SS_{xy} = -14408.672$	$\hat{\beta}_0 = 14{,}012.16711$
$\sum y = 126{,}942$	$SS_{yy} = 371{,}283{,}046$	$SSE = 345{,}594{,}779.6$
$\sum x^2 = 153.4123$	$\bar{x} = 3.112667$	$s^2 = 26{,}584{,}213.82$
$\sum y^2 = 1{,}445{,}567{,}804$	$\bar{y} = 8462.8$	$s = 5155.98815$
$\sum xy = 380{,}719.46$		

The least squares line is $\hat{y} = 14{,}012.17 - 1782.83x$.

Since we want to determine if there is an inverse linear relationship between raise and rating, we test:

$$H_0: \beta_1 = 0$$
$$H_a: \beta_1 < 0$$

Test statistic: $$t = \frac{\hat{\beta}_1}{s/\sqrt{SS_{xx}}} = \frac{-1782.83}{5155.988/\sqrt{8.08189}} = -.98$$

Rejection region: $\alpha = .05$, df $= n - 2 = 13$, $t_{.05} = 1.771$
Reject H_0 if $t < -1.771$

Conclusion: Do not reject H_0 at $\alpha = .05$. There is insufficient evidence of an inverse linear relationship between raise and rating.

b.

$n = 14$	$SS_{xx} = 6.3062929$	$\hat{\beta}_1 = -3886.726636$
$\sum x = 42.29$	$SS_{xy} = -24{,}510.8364$	$\hat{\beta}_0 = 19{,}680.33353$
$\sum y = 111{,}155$	$SS_{yy} = 313{,}807{,}433.2$	$SSE = 218{,}540{,}512.5$
$\sum x^2 = 134.0523$	$\bar{x} = 3.020714$	$s^2 = 18{,}211{,}709.38$
$\sum y^2 = 1{,}196{,}338{,}435$	$\bar{y} = 7939.642857$	$s = 4267.5179$
$\sum xy = 311{,}256.66$		

The least squares line is $\hat{y} = 19{,}680.33 - 3886.73x$.

Since we want to determine if there is an inverse linear relationship between raise and rating, we test:

H_0: $\beta_1 = 0$
H_a: $\beta_1 < 0$

Test statistic: $t = \dfrac{\hat{\beta}_1}{s/\sqrt{SS_{xx}}} = \dfrac{-3886.73}{4267.5179/\sqrt{6.30629}} = -2.29$

Rejection region: $\alpha = .05$, df $= n - 2 = 12$, $t_{.05} = 1.782$
Reject H_0 if $t < -1.782$

Conclusion: Reject H_0 at $\alpha = .05$. There is sufficient evidence of an inverse linear relationship between raise and rating.

c. Since we only know the relationship between raise, y, and rating, x, within the observed range of values of rating, it would be very risky to use this least squares line for the data in the table. The observed range of values for rating, x, is 1.52 to 4.11. The table contains estimates for raises for two values of x below 1.52 and two values of x greater than 4.11. We cannot be sure that the observed relationship between raise and rating holds outside the observed range.

d. As stated in the problem, typically only those with axes to grind respond to surveys. Thus, the reported average ratings may be quite different than the average ratings of the faculty as a whole. With such a low response rate, the chance that the sample is representative of the entire population is probably quite small. The results should be viewed with caution.

e. Because of the small response rate and the probable biased responses, the claim of the UFF should be viewed with skepticism.

3.71 Summary calculations yield:

$\sum x = .958 \quad \sum x^2 = .061638 \quad \sum xy = 10.2915$

$\sum y = 190.5 \quad \sum y^2 = 1978.33$

$$SS_{xy} = \sum xy - \frac{\sum x \sum y}{n} = 10.2915 - \frac{(.958)(190.5)}{20} = 1.16655$$

$$SS_{xx} = \sum x^2 - \frac{(\sum x)^2}{n} = .061638 - \frac{.958^2}{20} = .0157498$$

$$SS_{yy} = \sum y^2 - \frac{(\sum y)^2}{n} = 1978.33 - \frac{190.5^2}{20} = 163.8175$$

a.

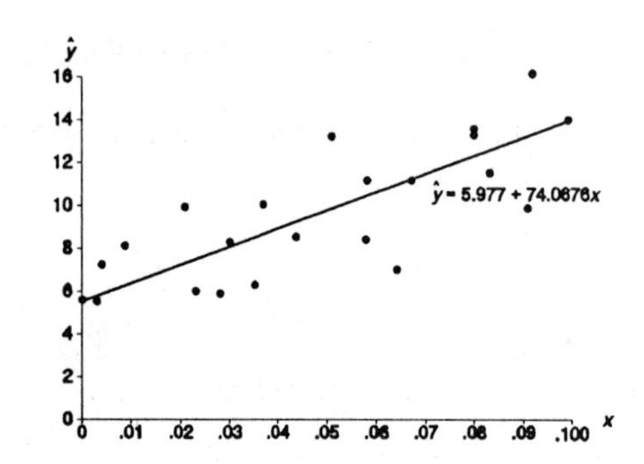

b. $\hat{\beta}_1 = \dfrac{SS_{xy}}{SS_{yy}} = \dfrac{1.16655}{.0157498} = 74.06760721 \approx 74.0676$

$$\hat{\beta}_0 = \bar{y} - \hat{\beta}_1 \bar{x} = \left(\frac{\sum y}{n}\right) - \hat{\beta}_1 \frac{\left(\sum x\right)}{n} = \frac{190.5}{20} - 74.06760721\left(\frac{.958}{20}\right) = 5.977$$

The least squares equation is $\hat{y} = 5.977 + 74.0676x$.

c. For $x = 0$, $\hat{y} = 5.977 + 74.0676(0) = 5.977$

For $x = .1$, $\hat{y} = 5.977 + 74.0676(.1) = 13.38376$

Connecting these points will yield the least squares line. See graph in part (a).

d. When $x = \$.070$, $\hat{y} = 5.977 + 74.0676(.070) = 11.161732$

e. $SSE = SS_{yy} - \hat{\beta}_1 SS_{xy} = 163.8175 - 74.06760721(1.16655) = 77.41393281$

$s^2 = \dfrac{SSE}{n - 2} = \dfrac{77.41393281}{20 - 2} = 4.300774 \qquad s = \sqrt{s^2} = 2.074$

The form of the prediction interval for y is:

$$\hat{y} = t_{\alpha/2}\, s \sqrt{1 + \frac{1}{n} + \frac{(x - \bar{x})^2}{SS_{xx}}} \quad \text{where } t_{.05} = 1.734 \text{ with df} = n - 1 = 18$$

$\hat{y} = 5.977 + 74.0676(.070) = 11.16$

The 90% prediction interval is:

$$11.16 \pm 1.734(2.074)\sqrt{1 + \frac{1}{20} + \frac{(.070 - .0479)^2}{.0157498}} \Rightarrow 11.16 \pm 3.74$$

$$\Rightarrow (7.42, 14.90)$$

f. The form of the confidence interval for $E(y)$ is:

$$\hat{y} \pm t_{\alpha/2}\, s \sqrt{\frac{1}{n} + \frac{(x - \bar{x})^2}{SS_{xx}}} \Rightarrow 11.16 \pm 1.734(2.074)\sqrt{\frac{1}{20} + \frac{(.070 - .0479)^2}{.0157498}}$$

$$\Rightarrow 11.16 \pm 1.02 \Rightarrow (10.14, 12.18)$$

g. The confidence interval for $E(y)$ is much narrower than the confidence interval for y.

h. Since the sample size would affect both n and SS_{xx}, increasing the sample size will reduce the widths of both intervals.

3.73 a. H_0: $\rho = 0$
H_a: $\rho \neq 0$

where ρ is the correlation coefficient for the population of passenger cars.

Test statistic: $t = \frac{r\sqrt{n-2}}{\sqrt{1-r^2}} = \frac{\sqrt{.676}\sqrt{n-2}}{\sqrt{1-.676}} = 1.444\sqrt{n-2}$

(We do not know n, so the exact value of t cannot be found.)

Conclusion: A p-value $< .05$ implies that we can reject H_0 for any α greater than or equal to .05. Since we have chosen $\alpha = .05$, there is sufficient evidence to reject H_0 at that level. There is sufficient evidence to indicate that x, distance between locations, is linearly related to y, urban travel time, for passenger cars.

b. H_0: $\rho = 0$
H_a: $\rho \neq 0$

where ρ is the correlation coefficient for the population of trucks.

Test statistic: $t = \frac{r\sqrt{n-2}}{\sqrt{1-r^2}} = \frac{\sqrt{.758}\sqrt{n-2}}{\sqrt{1-.758}} = 1.7698\sqrt{n-2}$

(We do not know n, so the exact value of t cannot be found.)

Conclusion: A p-value $< .01$ implies that we can reject H_0 for any α greater than or equal to .01. Since we have chosen $\alpha = .01$, there is sufficient evidence to reject H_0 at that level. There is sufficient evidence to indicate that x, distance between locations, is linearly related to y, urban travel time, for passenger cars.

c. The independent variable x, distance between locations, explains 67.6% and 75.8% of the variability in the dependent variable y, urban travel time, for passenger cars and trucks, respectively, when we use the least squares equation to predict y.

d. The estimated mean urban travel time for passenger cars when $x = 3$ is:

$$\hat{y} = 2.50 + 1.93(3) = 8.29 \text{ minutes}$$

e. The predicted urban travel time for a truck when $x = 5$ is:

$$\hat{y} = 1.85 + 3.86(5) = 21.15 \text{ minutes}$$

f. A measure of reliability for the expected urban travel time for passenger cars in part (d) would be to set up a confidence interval using the following formula:

$$\hat{y} \pm t_{\alpha/2}s\sqrt{\frac{1}{n} + \frac{(x_p - \bar{x})^2}{SS_{xx}}} \quad \text{where } x_p = 3$$

For the predicted urban travel time for a truck, the prediction interval would use the following formula:

$$\hat{y} \pm t_{\alpha/2} s \sqrt{1 + \frac{1}{n} + \frac{(x_p - \bar{x})^2}{SS_{xx}}} \quad \text{where } x_p = 5$$

Note: Though a significant linear relationship probably exists between these variables, the scattergram indicates that the relationship may be better described using a curve. This topic will be covered in later chapters.

3.75 a. The independent variable, x, the number of unique operands in the module, explains 74% of the variability in the dependent variable, y, the number of module defects, when we use the least squares equation to predict y.

b. H_0: $\rho = 0$
H_a: $\rho \neq 0$

We cannot compute the test statistic or determine the rejection region because n is not given. Because $r^2 = .74$ is fairly close to 1, there probably is a linear relationship between the number of module defects and the number of unique operands.

4 Multiple Regression

4.1 Associated with each independent variable in a multiple regression model is a β-coefficient. Since these coefficients are unknown, we estimate them using data from a sample of size n. Each estimate eliminates one degree of freedom available for estimating σ^2. Hence, if there are k independent variables in the model plus the y-intercept, β_0, then there will be $n - (k + 1)$ degrees of freedom left to estimate σ^2.

4.3 Let y = score on Zung scale
x_1 = score on ASO scale
x_2 = score on Burns scale
x_3 = score on Rotter scale

a. $E(y) = \beta_0 + \beta_1 x_1 + \beta_2 x_2 + \beta_3 x_3$

b. As with all p-values, we can reject H_0 if we are testing at $\alpha > p$-value. For $\alpha > .87$, we would reject H_0. For $\alpha < .87$, we would fail to reject H_0. In most situations, we will fail to reject H_0 where H_0: $\beta_2 = 0$.

4.5 If all else is held constant in the stand, allowing the age (x_1) of the dominant trees in the stand to mature by one year will increase the scenic beauty, y. This is the definition of the parameter estimate for β_1. Since the scenic beauty increases, the parameter estimate is positive.

4.7 a. **For Accountants:**

To determine if the model is adequate, we test:

H_0: $\beta_1 = \beta_2 = 0$
H_a: At least one $\beta_i \neq 0$

Test statistic:

$$F = \left[\frac{R^2}{1 - R^2}\right]\left[\frac{n - (k + 1)}{k}\right] = \left[\frac{.114}{1 - .114}\right]\left[\frac{169 - (2 + 1)}{2}\right] = 10.68$$

Rejection region: $\alpha = .05$, $\nu_1 = 2$, $\nu_2 = 166$, $F_{.05} \approx 3.07$
Reject H_0 if $F > 3.07$

Conclusion: Reject H_0. There is sufficient evidence to indicate the model is adequate at $\alpha = .05$.

For Truck Drivers:

To determine if the model is adequate, we test:

H_0: $\beta_1 = \beta_2 = 0$
H_a: At least one $\beta_i \neq 0$

Test statistic:

$$F = \left[\frac{R^2}{1 - R^2}\right]\left[\frac{n - (k + 1)}{k}\right] = \left[\frac{.298}{1 - .298}\right]\left[\frac{107 - (2 + 1)}{2}\right] = 22.07$$

Rejection region: $\alpha = .05$, $\nu_1 = 2$, $\nu_2 = 104$, $F_{.05} \approx 3.10$.
Reject H_0 if $F > 3.10$

Conclusion: Reject H_0. There is sufficient evidence to indicate the model is adequate at $\alpha = .05$.

b. **For Accountants:**

$\hat{\beta}_1 = -1.40$ is an estimate of the amount the curve is shifted along the x-axis. This does not have a practical interpretation.

$\hat{\beta}_2 = 1.13$ is an estimate of the amount of curvature in the least squares curved line. The sign of $\hat{\beta}_2$ is positive which implies upward curvature in the probability of turnover-performance rating relationship. This has a practical interpretation.

For Truck Drivers:

$\hat{\beta}_1 = -1.50$ is an estimate of the amount the curve is shifted along the x-axis. This does not have a practical interpretation.

$\hat{\beta}_2 = 1.22$ is an estimate of the amount of curvature in the least squares curved line. The sign of $\hat{\beta}_2$ is positive which implies an upward curvature in the probability of turnover-performance rating relationship. This has a practical interpretation.

c. To determine if there is an upward curvature in the relationship between turnover and performance, we test:

H_0: $\beta_2 = 0$
H_a: $\beta_2 > 0$

Test statistic: $t = 3.23$

Rejection region: $\alpha = .05$, df $= 166$, $t_{.05} = 1.645$
Reject H_0 if $t > 1.645$

Conclusion: Reject H_0. There is sufficient evidence to indicate an upward curvature in the relationship between turnover and performance at $\alpha = .05$.

d. To determine if there is an upward curvature in the relationship between turnover and performance, we test:

H_0: $\beta_2 = 0$
H_a: $\beta_2 > 0$

Test statistic: $t = 4.70$

Rejection region: $\alpha = .05$, df = 104, $t_{.05} = 1.660$
Reject H_0 if $t > 1.660$

Conclusion: Reject H_0. There is sufficient evidence to indicate an upward curvature in the relationship between turnover and performance at $\alpha = .05$.

4.9 a. $E(y) = \beta_0 + \beta_1 x_1 + \beta_2 x_1^2$

b. If wages (y) increase with education (x_1), but at a decreasing rate, then the coefficient corresponding to the x_1^2 term will be negative. Thus, the test would be:

H_0: $\beta_2 = 0$
H_a: $\beta_2 < 0$

4.11 a. H_0: $\beta_1 = \beta_2 = \cdots = \beta_{18} = 0$
H_a: At least one of the coefficients is nonzero

Test statistic: $F = \dfrac{R^2/k}{(1 - R^2)/[n - (k + 1)]} = \dfrac{.95/18}{.05/1} = 1.0556$

Rejection region: $\alpha = .05$, $\nu_1 = 18$, $\nu_2 = 1$, $F_{.05} = 247$
Reject H_0 if $F > 247$

Conclusion: There is insufficient evidence to reject H_0 at $\alpha = .05$. What happened? The inclusion of 18 independent variables in the model reduces the number of degrees of freedom in the denominator to 1. The result is an inflation of the critical value, which makes it difficult to reject H_0. In order to have more degrees of freedom available for estimating σ^2, the researcher should either collect more data or include fewer independent variables in the model.

b. $R_a = 1 - \dfrac{(n - 1)}{n - (k + 1)}(1 - R^2) = 1 - \dfrac{20 - 1}{20 - (18 + 1)}(1 - .95) = .05$

After adjusting for the sample size and the number of parameters in the model, approximately 5% of the sample variation in the GNP can be "explained" by the model with 18 independent variables.

4.13 a. The estimate of the β parameter corresponding to the independent variable, rental price, is 2.87. For each unit increase in rental price, the mean number of homeless is estimated to increase by 2.87, all other variables held constant.

b. To determine if homelessness decreases as employment growth increases, we test:

H_0: $\beta_4 = 0$
H_a: $\beta_4 < 0$

Test statistic: $t = -2.71$ (from printout).

Rejection region: $\alpha = .05$, df = 33, $t_{.05} \approx 1.697$
Reject H_0 if $t < -1.697$

Conclusion: Reject H_0. There is sufficient evidence to indicate that homelessness decreases as employment growth increases at $\alpha = .05$.

c. For all cases, the hypotheses are:

H_0: $\beta_i = 0$
H_a: $\beta_i \neq 0$

Rejection region: $\alpha = .05$, df = 33, $t_{.025} \approx 2.042$
Reject H_0 if $t < -2.042$ or $t > 2.042$

Based on the rejection region and the printed t-values, the following variables are significantly related to homelessness:

Rental price, Employment growth, AFDC benefits, and SSI benefits

d. The tests performed in part (c) are not independent. Thus, the α level is inflated. This means that the chance of rejecting at least one H_0 when it is true is greater than $\alpha = .05$.

e. $R_a^2 = .83$. This means that 83% of the sample variation in homelessness rates is explained by the model containing the 16 independent variables, adjusting for the sample size and the number of β's in the model.

4.15 a. H_0: $\beta_1 = \beta_2 = \beta_3 = 0$
H_a: At least one of the coefficients is nonzero

Test statistic: $F = 9.6893$

Rejection region: $\alpha = .10$, $\nu_1 = 3$, $\nu_2 = 286$, $F_{.10} \approx 2.08$
Reject H_0 if $F > 2.08$

Conclusion: Reject H_0 at $\alpha = .10$. There is sufficient evidence that the model is adequate for predicting coupon redemption rate.

b. If coupon users are more price-conscious than nonusers, then price-consciousness score and coupon redemption rate should be positively related. We test:

H_0: $\beta_1 = 0$
H_a: $\beta_1 > 0$

Test statistic: $t = 1.444$

Rejection region: $\alpha = .10$, df $= n - 4 = 286$, $t_{.10} = 1.282$
Reject H_0 if $t > 1.282$

Conclusion: Reject H_0 at $\alpha = .10$. There is sufficient evidence to indicate that coupon users are more price-conscious than nonusers.

c. We estimate that the mean coupon redemption rate will decrease by .13134 for each additional unit increase in time-value score, with price-consciousness score and satisfaction/pride score held constant. That is, as one values his/her time more, coupon redemption rate will decrease.

4.17 a. H_0: $\beta_1 = \beta_2 = \beta_3 = \beta_4 = 0$
H_a: At least one $\beta_i \neq 0$ for $i = 1, 2, 3, 4$

Test statistic: $F = 72.11$

Rejection region: α is not given, so we will use $\alpha = .05$; $\nu_1 = 4$, $\nu_2 = 31$, $F_{.05} \approx 2.69$. Reject H_0 if $F > 2.69$.

Also, the p-value is given as 0.000. Since this is smaller than any reasonable value of α, we will reject H_0.

Conclusion: Reject H_0. There is sufficient evidence to indicate the model is adequate at $\alpha = .05$.

b. From the printout, the 95% prediction interval is (47, 3825). We are 95% confident that the actual manhours needed to erect an industrial field will be between 47 and 3825 when the boiler capacity is 150,000 lb/hr, the design pressure is 500 psi, when using the steam drum type.

4.19 a. To determine if the model is useful for estimating mean bid price, we test:

H_0: $\beta_1 = \beta_2 = 0$
H_a: At least one $\beta_i \neq 0$, $i = 1, 2$

Test statistic: $F = 120.651$

The p-value $= .0001$. Since the p-value is less than $\alpha = .05$, H_0 is rejected. There is sufficient evidence to indicate the model is useful for estimating mean bid price at $\alpha = .05$.

b. To determine if the mean bid price increases as the number of bidders increases for road construction projects of the same length, we test:

H_0: $\beta_2 = 0$
H_a: $\beta_2 > 0$

Test statistic: $t = 9.86$

The p-value $= .0001/2 = .00005$. Since the p-value is less than $\alpha = .01$, H_0 is rejected. There is sufficient evidence to indicate the mean bid price increases as the number of bidders increases for road construction projects of the same length at $\alpha = .01$.

c. The 95% confidence interval is (253.81, 821.39). We are 95% confident the mean bid price will be between $253,810 and $821,390 when there are 100 road miles and 7 bidders.

4.21 a. Let x_1 and x_2 represent the two quantitative independent variables. A first-order linear model includes only the first-order terms for the variables:

$$E(y) = \beta_0 + \beta_1 x_1 + \beta_2 x_2$$

b. Let x_1, x_2, x_3, and x_4 represent the four quantitative independent variables. A first-order linear model includes only the first-order terms for the variables:

$$E(y) = \beta_0 + \beta_1 x_1 + \beta_2 x_2 + \beta_3 x_3 + \beta_4 x_4$$

4.23 a. $E(y) = \beta_0 + \beta_1 x$, where $x = \begin{cases} 1 \text{ if A} \\ 0 \text{ if B} \end{cases}$

β_0 = mean of y for the $x = B$ level $= \mu_B$
β_1 = difference in the mean levels of y for the x = A and x = B levels $= \mu_A - \mu_B$

b. For a qualitative independent variable with four levels (A, B, C, and D), the model requires three dummy variables as shown below. (We have arbitrarily selected level D as the "base" level.)

$x_1 = \begin{cases} 1 & \text{if level A} \\ 0 & \text{if not} \end{cases}$ $\quad x_2 = \begin{cases} 1 & \text{if level B} \\ 0 & \text{if not} \end{cases}$ $\quad x_3 = \begin{cases} 1 & \text{if level C} \\ 0 & \text{if not} \end{cases}$

Then the model is written $E(y) = \beta_0 + \beta_1 x_1 + \beta_2 x_2 + \beta_3 x_3$

The following table shows the values of the dummy variables and the mean response $E(y)$ for each of the four levels.

Level	x_1	x_2	x_3	$E(y)$
A	1	0	0	$\beta_0 + \beta_1 = \mu_A$
B	0	1	0	$\beta_0 + \beta_2 = \mu_B$
C	0	0	1	$\beta_0 + \beta_3 = \mu_C$
D	0	0	0	$\beta_0 = \mu_D$

From the table, we obtain the interpretations of the β's.

$$\beta_0 = \mu_D,\ \beta_1 = \mu_A - \mu_D,\ \beta_2 = \mu_B - \mu_D, \text{ and } \beta_3 = \mu_C - \mu_D$$

4.25 a. $y = 1 + 2x_1 + (1) - 3(3)$
$y = -7 + 2x_1$
For $x_1 = 0$, $y = -7$
For $x_1 = 6$, $y = 5$

Two points determine the line.

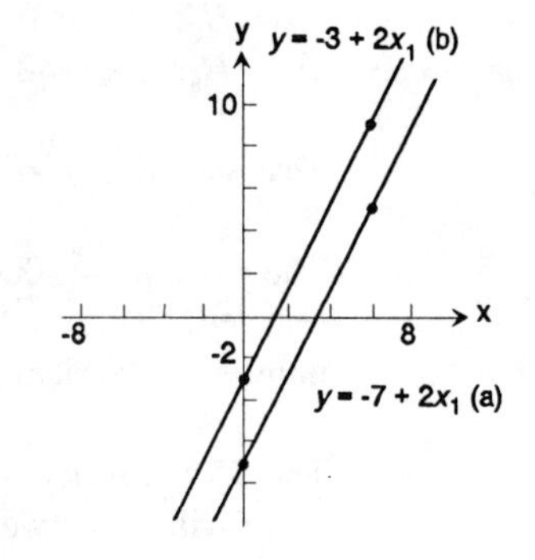

b. 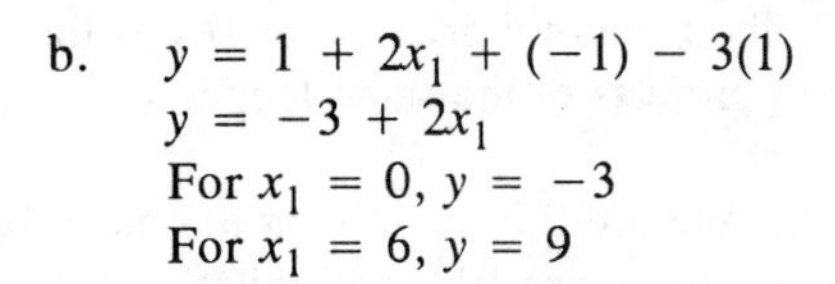
$y = 1 + 2x_1 + (-1) - 3(1)$
$y = -3 + 2x_1$
For $x_1 = 0$, $y = -3$
For $x_1 = 6$, $y = 9$

Two points determine the line.

c. The geometric relationship in a first-order model for $E(y)$ as a function of one independent variable for various combinations of values of the other independent variables is parallel lines. The y-intercept is determined by the combination of values of the other independent variables. The slope is the coefficient of the one independent variable.

4.27 a. With $x_2 = 0$, $E(y) = 1 + x_1 - (0) + x_1(0) + 2x_1^2 + (0)^2 = 1 + x_1 + 2x_1^2$

For $x_1 = 0$, $E(y) = 1$ $\quad$ For $x_1 = 1$, $E(y) = 4$
For $x_1 = 2$, $E(y) = 11$ $\quad$ For $x_1 = -1$, $E(y) = 2$
For $x_1 = -2$, $E(y) = 7$

With $x_2 = 1$, $E(y) = 1 + x_1 - (1) + x_1(1) + 2x_1^2 + (1)^2 = 1 + 2x_1 + 2x_1^2$

For $x_1 = 0$, $E(y) = 1$ For $x_1 = 1$, $E(y) = 5$
For $x_1 = 2$, $E(y) = 13$ For $x_1 = -1$, $E(y) = 1$
For $x_1 = -2$, $E(y) = 5$

With $x_2 = 2$, $E(y) = 1 + x_1 - (2) + x_1(2) + 2x_1^2 + (2)^2 = 3 + 3x_1 + 2x_1^2$

For $x_1 = 0$, $E(y) = 3$ For $x_1 = 1$, $E(y) = 8$
For $x_1 = 2$, $E(y) = 17$ For $x_1 = -1$, $E(y) = 2$
For $x_1 = -2$, $E(y) = 5$

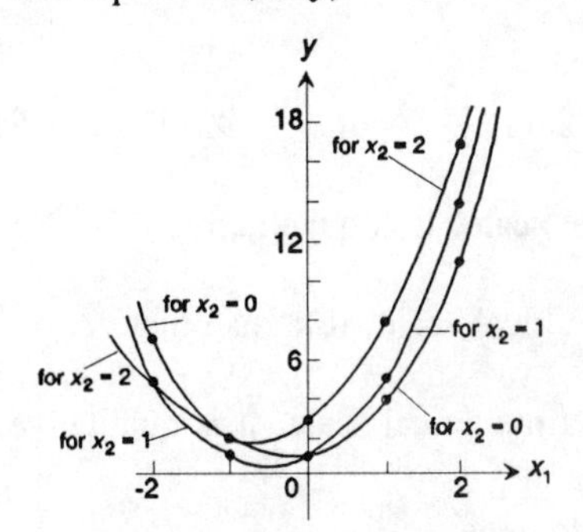

b. The curves are second-order because they contain a squared term (x_1^2).

c. They are overlapping (intersecting) parabolas.

d. Yes, the shape of the x_1 curve changes for different x_2 values.

e. It changes the shape of the curve for different x_2 values by shifting the curve along the x_1 axis.

4.29 a. The least squares prediction equation is:

$$\hat{y} = 64.82 + 1.05x_1 - 10.53x_2 + .27x_3 + 9.46x_4 - 92.97x_5$$

b. $\hat{\beta}_1 = 1.05$. The mean annual health care expenditures is estimated to increase by \$1.05 for each additional year of age, all other variables held constant.

$\hat{\beta}_4 = 9.46$. The mean annual health care expenditures is estimated to increase by \$9.46 for each additional concomitant maintenance medication, all other variables held constant.

c. All terms in the model with p-values less than .05 are significant. Those variables whose p-values are not less than .05 are not significant. From the table, gender and race are not significant factors in this model.

d. The F value given is $F = 37.84$. This value is used to test the overall adequacy of the model:

H_0: $\beta_1 = \beta_2 = \beta_3 = \beta_4 = \beta_5 = 0$
H_a: At least one $\beta_i \neq 0$ for $i = 1, 2, 3, 4, 5$

Test statistic: $F = 37.84$

Rejection region: $\alpha = .05$, $\nu_1 = 5$, $\nu_2 = 282 - 6 = 276$, $F_{.05} \approx 2.21$
Reject H_0 if $F > 2.21$

Conclusion: Reject H_0. There is sufficient evidence that the model is adequate for predicting annual health care expenditures at $\alpha = .05$.

e. $R^2 = .4357$. 43.57% of the variability in the annual health care expenditures is explained by the model containing the variables age, gender, race, regimen, and health education.

f. For $x_1 = 45$, $x_2 = 0$, $x_3 = 1$, $x_4 = 3$, and $x_5 = 0$,

$$\hat{y} = 64.82 + 1.05(45) - 10.53(0) + .27(1) + 9.46(3) - 92.97(0) = \$140.72.$$

4.31 a. Since temperature is measured on a numerical scale, it is quantitative.

b. Since relative humidity is measured on a numerical scale, it is quantitative.

c. Since organic compound is not measured on a numerical scale, it is qualitative.

d. Let $x_1 = \begin{cases} 1 \text{ if benzene} \\ 0 \text{ if not} \end{cases}$ $x_2 = \begin{cases} 1 \text{ if toulene} \\ 0 \text{ if not} \end{cases}$

$x_3 = \begin{cases} 1 \text{ if chloroform} \\ 0 \text{ if not} \end{cases}$ $x_4 = \begin{cases} 1 \text{ if methanol} \\ 0 \text{ if not} \end{cases}$

The model would then be:

$$E(y) = \beta_0 + \beta_1 x_1 + \beta_2 x_2 + \beta_3 x_3 + \beta_4 x_4$$

e. β_0 = mean sorption of anisole vapor on clay minerals, μ_A
β_1 = difference in mean sorption on clay minerals between benzene and anisole, $\mu_B - \mu_A$
β_2 = difference in mean sorption on clay minerals between toluene and anisole, $\mu_T - \mu_A$
β_3 = difference in mean sorption on clay minerals between chloroform and anisole, $\mu_C - \mu_A$
β_4 = difference in mean sorption on clay minerals between methanol and anisole, $\mu_M - \mu_A$

f. To see if the mean retention coefficients of the five organic compounds differ, we test:

H_0: $\beta_1 = \beta_2 = \beta_3 = \beta_4 = 0$
H_a: At least one $\beta_i \neq 0$ for $i = 1, 2, 3, 4$

The test statistic would be an F.

4.33 To determine if the model is useful, we test:

H_0: $\beta_1 = \beta_2 = \beta_3 = \beta_4 = 0$
H_a: At least one $\beta_i \neq 0$ for $i = 1, 2, 3, 4$

Test statistic: $$F = \frac{R^2/k}{(1 - R^2)/[n - (k + 1)]} = \frac{.742/4}{(1 - .742)/[103 - (4 + 1)]} = 70.46$$

Rejection region: $\alpha = .05$, $\nu_1 = 4$, $\nu_2 = 98$, $F_{.05} \approx 2.45$
Reject H_0 if $F > 2.45$

Conclusion: Reject H_0. There is sufficient evidence (at $\alpha = .05$) to indicate the model is useful for predicting the TCDD level.

$R^2 = .742$. 74.2% of the variability in the TCDD levels is explained by the model containing logarithm of years of exposure, number of years since last exposure, age, and body mass.

Since all of the p-values associated with independent variables are less than .001, all of the variables included in the model are significant.

4.35 a. H_0: $\beta_1 = \beta_2 = \beta_3 = \beta_4 = 0$
H_a: At least one coefficient is nonzero

Test statistic: $F = \dfrac{R^2/k}{(1 - R^2)/[n - (k + 1)]} = \dfrac{.95/4}{.05/823} = 3909.25$

Rejection region: $\alpha = .05$, $\nu_1 = 4$, $\nu_2 = 823$, $F_{.05} \approx 2.37$
Reject H_0 if $F > 2.37$

Conclusion: Reject H_0. There is sufficient evidence (at $\alpha = .05$) to indicate that the model is useful for predicting the natural log of per capita consumption of cigarettes by persons of smoking age.

b. The standard deviation, s, is the estimate of σ, the standard deviation of ϵ. Most of the observed y-values should lie within $2s$ or $2(.047) = .094$ of their least squares predicted values, $\hat{y}$.

c. H_0: $\beta_4 = 0$
H_a: $\beta_4 < 0$

d. A value of .033 for $\hat{\beta}_4$ implies that, for every one unit increase in per capita index of advertising expenditures, the mean natural log of per capita consumption of cigarettes increases by .033, all other variables held constant.

e. No, to reject H_0 in favor of a left-tailed H_a, the test statistic would need to be negative. A positive value for $\hat{\beta}_4$ would never lead to a negative test statistic; therefore, H_0 would not be rejected.

4.37 Models (a) and (b) are nested. The complete model is $E(y) = \beta_0 + \beta_1x_1 + \beta_2x_2$ and the reduced model is $E(y) = \beta_0 + \beta_1x_1$.

Models (a) and (d) are nested. The complete model is $E(y) = \beta_0 + \beta_1x_1 + \beta_2x_2 + \beta_3x_1x_2$ and the reduced model is $E(y) = \beta_0 + \beta_1x_1 + \beta_2x_2$.

Models (a) and (e) are nested. The complete model is $E(y) = \beta_0 + \beta_1x_1 + \beta_2x_2 + \beta_3x_1x_2 + \beta_4x_1^2 + \beta_5x_2^2$ and the reduced model is $E(y) = \beta_0 + \beta_1x_1 + \beta_2x_2$.

Models (b) and (c) are nested. The complete model is $E(y) = \beta_0 + \beta_1x_1 + \beta_2x_1^2$ and the reduced model is $E(y) = \beta_0 + \beta_1x_1$.

Models (b) and (d) are nested. The complete model is $E(y) = \beta_0 + \beta_1x_1 + \beta_2x_2 + \beta_3x_1x_2$ and the reduced model is $E(y) = \beta_0 + \beta_1x_1$.

Models (b) and (e) are nested. The complete model is $E(y) = \beta_0 + \beta_1x_1 + \beta_2x_2 + \beta_3x_1x_2 + \beta_4x_1^2 + \beta_5x_2^2$ and the reduced model is $E(y) = \beta_0 + \beta_1x_1$.

Models (c) and (e) are nested. The complete model is $E(y) = \beta_0 + \beta_1x_1 + \beta_2x_2 + \beta_3x_1x_2 + \beta_4x_1^2 + \beta_5x_2^2$ and the reduced model is $E(y) = \beta_0 + \beta_1x_1 + \beta_4x_1^2$.

Models (d) and (e) are nested. The complete model is $E(y) = \beta_0 + \beta_1x_1 + \beta_2x_2 + \beta_3x_1x_2 + \beta_4x_1^2 + \beta_5x_2^2$ and the reduced model is $E(y) = \beta_0 + \beta_1x_1 + \beta_2x_2 + \beta_3x_1x_2$.

4.39 a. H_0: $\beta_5 = \beta_6 = \beta_7 = \beta_8 = 0$
H_a: At least one coefficient is not 0

For males:

Rejection region: $\alpha = .05$, $\nu_1 = 4$, $\nu_2 = n - (k + 1) = 235$, $F_{.05} \approx 2.37$
Reject H_0 if $F > 2.37$

For females:

Rejection region: $\alpha = .05$, $\nu_1 = 4$, $\nu_2 = n - (k + 1) = 144$, $F_{.05} \approx 2.45$
Reject H_0 if $F > 2.45$

b. For males, reduced model: $R^2 = .218$. This implies that 21.8% of the sample variability of the intrinsic job satisfaction scores is explained by the model containing age, education level, firm experience, and sales experience.

For males, complete model: $R^2 = .408$. This implies that 40.8% of the sample variability of the intrinsic job satisfaction scores is explained by the model containing age, education level, firm experience, sales experience, contingent reward behavior, noncontingent reward behavior, contingent punishment behavior, and noncontingent punishment behavior.

Since the R^2 value increased from .218 to .408 after the 4 supervisory behavior variables were added to the model, it appears that they did have an impact on intrinsic job satisfaction for the males.

For females, reduced model: $R^2 = .268$. This implies that 26.8% of the sample variability of the intrinsic job satisfaction scores is explained by the model containing age, education level, firm experience, and sales experience.

For females, complete model: $R^2 = .496$. This implies that 49.6% of the sample variability of the intrinsic job satisfaction scores is explained by the model containing age, education level, firm experience, sales experience, contingent reward behavior, noncontingent reward behavior, contingent punishment behavior, and noncontingent punishment behavior.

Since the R^2 value increased from .268 to .496 after the 4 supervisory behavior variables were added to the model, it appears that they did have an impact on intrinsic job satisfaction for the females.

c. Test statistic: $F_{\text{males}} = 13.00$ $F_{\text{females}} = 9.05$

Rejection region:

Males: $\alpha = .05$, $\nu_1 = 4$, $\nu_2 = 235$, $F_{.05} \approx 2.37$
Reject H_0 if $F > 2.37$

Females: $\alpha = .05$, $\nu_1 = 4$, $\nu_2 = 144$, $F_{.05} \approx 2.45$
Reject H_0 if $F > 2.45$

Conclusion: For both males and females, reject H_0 at $\alpha = .05$. There is sufficient evidence to indicate that at least one of the four supervisory behavior variables affect intrinsic job satisfaction.

4.41 a. We would test:

H_0: $\beta_2 = 0$
H_a: $\beta_2 \neq 0$

b. Test statistic: $F = \dfrac{(\text{SSE}_R - \text{SSE}_C)/(k - g)}{\text{SSE}_C/[n - (k + 1)]} = \dfrac{(26.01 - 25.44)/(2 - 1)}{25.44/[32 - (2 + 1)]} = \dfrac{.57}{.8772} = .65$

Rejection region: $\alpha = .05$, $\nu_1 = k - g = 1$, $\nu_2 = n - (k + 1) = 29$, $F_{.05} = 4.18$
Reject H_0 if $F > 4.18$

Conclusion: Do not reject H_0. There is insufficient evidence to indicate the addition of the aggression rating improves the predictive ability of the model at $\alpha = .05$.

4.43 a. H_0: $\beta_1 = \beta_2 = \beta_3 = \beta_4 = 0$
H_a: At least one $\beta_i \neq 0$, $i = 1, 2, 3, 4$

Test statistic: $F = \dfrac{R^2/k}{(1 - R^2)/[n - (k + 1)]} = \dfrac{.11/4}{.89/95} = 2.94$

Rejection region: $\alpha = .05$, $\nu_1 = 4$, $\nu_2 = 95$, $F_{.05} \approx 2.45$
Reject H_0 if $F > 2.45$

Conclusion: Reject H_0. There is sufficient evidence (at $\alpha = .05$) to indicate that the model is useful for predicting performance rating.

b. H_0: $\beta_5 = \beta_6 = 0$
H_a: At least one of the coefficients β_5 and β_6 differs from zero

Test statistic: $F = \dfrac{(\text{SSE}_R - \text{SSE}_C)/(k - g)}{\text{SSE}_C/[n - (k + 1)]} = \dfrac{(352 - 341)/(6 - 4)}{341/[100 - (6 + 1)]} = 1.50$

Rejection region: $\alpha = .05$, $\nu_1 = 2$, $\nu_2 = 93$, $F_{.05} \approx 3.07$
Reject H_0 if $F > 3.07$

Conclusion: Do not reject H_0. There is insufficient evidence (at $\alpha = .05$) to indicate that managerial level and subordinate-related behavior contribute additional information for the prediction of performance rating.

c. H_0: $\beta_7 = 0$
H_a: $\beta_7 \neq 0$

Test statistic: $F = \frac{(\text{SSE}_R - \text{SSE}_C)/(k - g)}{\text{SSE}_C/[n - (k + 1)]} = \frac{(341 - 321)/(7 - 6)}{321/[100 - (7 + 1)]} = 5.73$

Rejection region: $\alpha = .05$, $\nu_1 = 1$, $\nu_2 = 92$, $F_{.05} \approx 3.92$
Reject H_0 if $F > 3.92$

Conclusion: Reject H_0. There is sufficient evidence (at $\alpha = .05$) to indicate that the interaction between managerial level and subordinate-related behavior is important.

d. An interaction exists between variables x_5 and x_6, making the main effects difficult to interpret. The effect of subordinate-related managerial behavior score, x_6, on performance rating depends on whether the individual is a middle/upper-level manager or a lower level manager.

4.45 a. The variable with the most extreme test statistic would be declared the best one-variable predictor of y.

Independent Variable	$t = \frac{\hat{\beta}_i}{s_{\hat{\beta}_i}}$
x_1	3.81
x_2	−90.0
x_3	2.98
x_4	1.21
x_5	−6.03
x_6	.86

x_2 would be declared the best one-variable predictor of y.

b. Since x_2 would be significant at $\alpha = .05$, it would be entered in the model at this stage.

c. The next phase involves fitting all models of the form $E(y) = \beta_0 + \beta_1 x_i + \beta_2 x_2$, $i = 1, 3, 4, 5, 6$ to find the best two-predictor model for y.

4.47 a. We must find the rejection region for testing H_0: $\beta_i = 0$ vs H_a: $\beta_i \neq 0$, using $\alpha = .01$, $\text{df} = n - (k + 1) = 105 - (k + 1) = 104 - k$, where $k = 1, 2, \ldots, 8$. Thus, df ranges from 103 to 96. The most conservative value of t is $t_{.005} \approx 2.64$. Reject H_0 if $t < -2.64$ or $t > 2.64$.

Based on the above rejection region, the variables listed that would be used to model rural property values are:

Residential land, Seedlings and saplings, Percentage ponds, Distance to state park, Branches or springs, and Site index.

b. A complete second-order model would be:

$$E(y) = \beta_0 + \beta_1x_1 + \beta_2x_2 + \beta_3x_3 + \beta_4x_4 + \beta_5x_1x_2 + \beta_6x_1x_3 + \beta_7x_1x_4 + \beta_8x_2x_3 + \beta_9x_2x_4 + \beta_{10}x_3x_4 + \beta_{11}x_1^2 + \beta_{12}x_2^2 + \beta_{13}x_3^2 + \beta_{14}x_4^2 + \beta_{15}x_5 + \beta_{16}x_6 + \beta_{17}x_5x_6 + \text{all quantitative} \times \text{qualitative interaction terms (28 more terms)}$$

where x_1 = seedlings and saplings, x_2 = percent ponds

x_3 = distance to state park, x_4 = site index

$x_5 = \begin{cases} 1 & \text{if residential land,} \\ 0 & \text{otherwise} \end{cases}$ $x_6 = \begin{cases} 1 & \text{if branches or springs,} \\ & \text{otherwise} \end{cases}$

4.49 a. There are (i) $\begin{pmatrix} 4 \\ 1 \end{pmatrix} = \dfrac{4!}{1!3!} = 4$ models with 1 variable

(ii) $\begin{pmatrix} 4 \\ 2 \end{pmatrix} = \dfrac{4!}{2!2!} = 6$ models with 2 variables

(iii) $\begin{pmatrix} 4 \\ 3 \end{pmatrix} = \dfrac{4!}{3!1!} = 4$ models with 3 variables

and (iv) $\begin{pmatrix} 4 \\ 4 \end{pmatrix} = \dfrac{4!}{4!0!} = 1$ model with 4 variables

b. Using SAS,

(i) for the 1 variable model, the maximum $R^2 = .2130$, minimum MSE = 193.7990, the minimum $C_p = 2.5445$, and the minimum PRESS = 10,507.429196

(ii) for the 2 variable model, the maximum $R^2 = .2473$, minimum MSE = 189.1257, the minimum $C_p = 2.3391$, and the minimum PRESS = 10,460.991189

(iii) for the 3 variable model, the maximum $R^2 = .2663$, minimum MSE = 188.2066, the minimum $C_p = 3.1225$, and the minimum PRESS = 10,489.283249

(iv) for the 4 variable model, the $R^2 = .2682$, MSE = 191.7114, the $C_p = 5$, and PRESS = 10,710.356996

c.

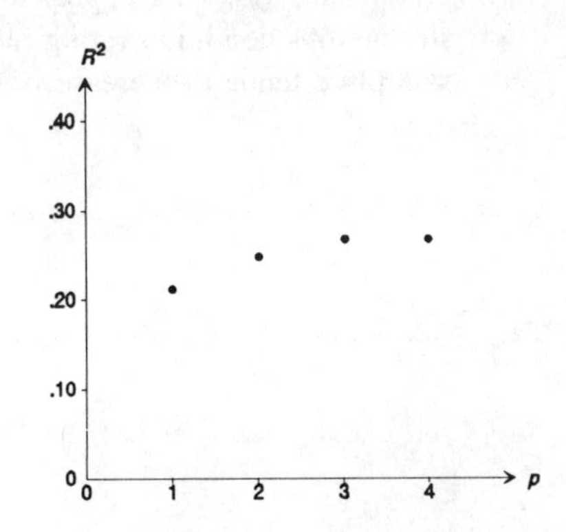

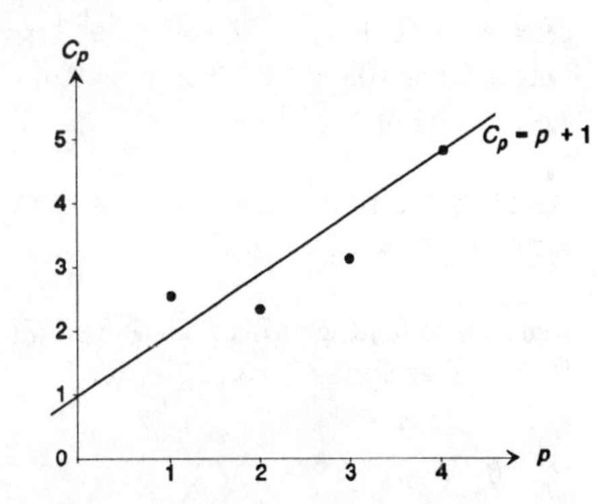

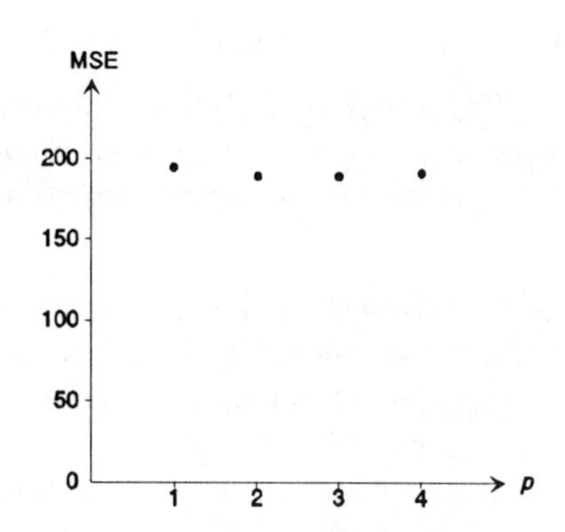

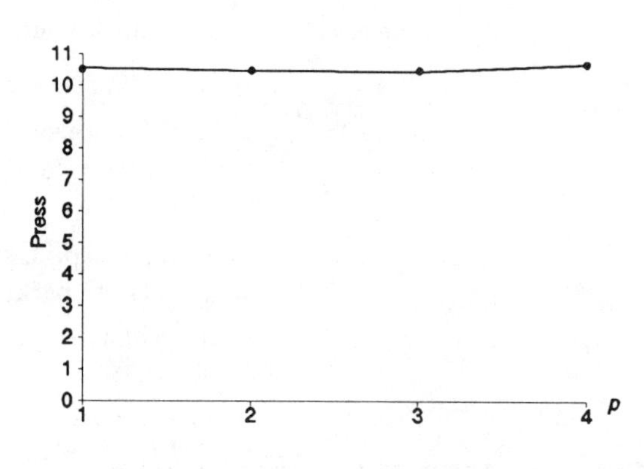

d. For the plots R^2 versus p, C_p versus p, MSE versus p, and Press versus p, it appears the best 1 variable model or the best 2 variable model is the best. For all values of p, the R^2 values are very similar, the C_p values are very similar, as are the MSE values and Press values. The best 1 variable model is the one with X3, while the best 2 variable model is the one with X3 and X2 or the one with X3 and X4 (from press value).

4.51 a.

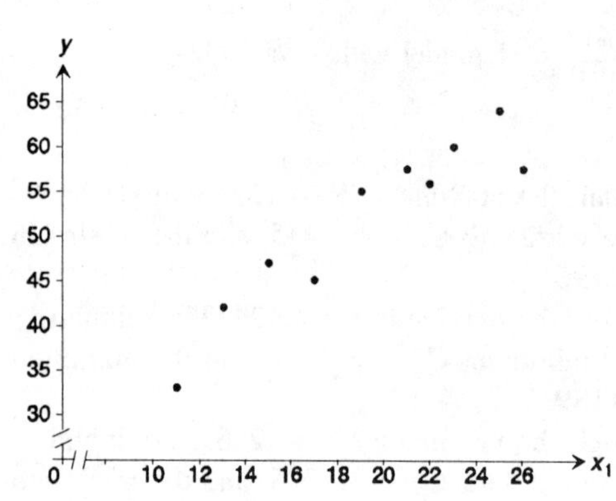

Yes, there appears to be a positive linear relationship between home size and sale price. As home size increases, sale price tends to increase.

b.

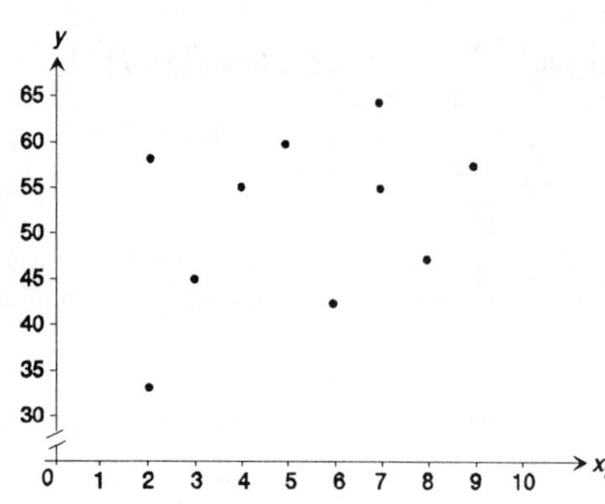

Yes, there appears to be a positive linear relationship between condition rating and sale price, but the relationship does not appear to be strong. As condition rating increases, sale price tends to increase.

c. H_0: $\beta_1 = 0$
H_a: $\beta_1 \neq 0$

Test statistic: $t = \dfrac{\hat{\beta}_1}{s_{\hat{\beta}_1}} = 24.561$ (This value was obtained directly from the printout.)

Rejection region: $\alpha = .01$, df $= n - 3 = 7$, $t_{.005} = 3.499$
Reject H_0 if $t < -3.499$ or $t > 3.499$

Conclusion: Reject H_0. There is sufficient evidence (at $\alpha = .01$) to indicate that home size and sale price are linearly related. Since $\hat{\beta}_1$ is a positive value, we can conclude that the relationship is positive.

Note: This same conclusion is more conveniently reached using the p-value obtained from the printout. A p-value of .0000 implies we will reject H_0 for any α greater than or equal to .0000. Since $\alpha = .01$, for this problem, we would reject H_0.

d. A 99% confidence interval for β_2 is $\hat{\beta}_2 \pm t_{.005} s_{\hat{\beta}_2}$ where $t_{.005} = 3.499$ is based on $n - 3 = 7$ df. Substitution yields:

$$1.278 \pm 3.499(.1444) \Rightarrow 1.278 \pm .505 \Rightarrow (.773, 1.783)$$

We are 99% confident that the change in the mean sale price for each unit change in condition rating is between .773 and 1.783 or between \$773 and \$1783, with home size held constant.

4.53 a. A 95% confidence interval for β_3 is $\hat{\beta}_3 \pm t_{.025} s_{\hat{\beta}_3}$ where $t_{.025} = 1.960$ is based on $n - 4 = 181$ df. Substitution yields:

$$-75.51 \pm 1.96(13.35) \Rightarrow -75.51 \pm 26.166 \Rightarrow (-101.676, -49.344)$$

We are 95% confident that the change in the mean percentage of vacant land zoned for residential use decreases from between -101.676 and -49.344 for each additional unit change in proportion of total tax base derived from nonresidential property, with proportion of existing land in nonresidential use remains constant.

b. H_0: $\beta_2 = 0$
H_a: $\beta_2 \neq 0$

Test statistic: $t = \dfrac{\hat{\beta}_2}{s_{\hat{\beta}_2}} = \dfrac{166.80}{120.88} = 1.38$

Rejection region: $\alpha = .05$, df $= n - 4 = 181$, $t_{.025} = 1.960$
Reject H_0 if $t < -1.96$ or $t > 1.96$

Conclusion: Do not reject H_0. There is insufficient evidence to indicate a curvilinear relationship exists between percentage of land zoned for residential use and proportion of existing land in nonresidential use.

c. The adjusted R^2 value is .25. 25% of the variability in the percentage of vacant land zoned for residential use is explained by the model containing the proportion of existing land in nonresidential use, this variable squared, and the proportion of total tax base derived from nonresidential property, adjusted for the sample size and the number of variables in the model.

d. To determine if the model is useful in predicting y, we test:

H_0: $\beta_1 = \beta_2 = \beta_3 = 0$
H_a: At least one $\beta_i \neq 0$ for $i = 1, 2, 3$

Test statistic: $F = 21.86$

Since the p-value associated with this test statistic is $p < .01$, H_0 is rejected. There is sufficient evidence to indicate that the model is statistically useful for predicting y for $\alpha > .01$.

4.55 a. H_0: $\beta_1 = \beta_2 = \beta_3 = 0$
H_a: At least one of the coefficients is nonzero

Test statistic: $F = \dfrac{R^2/k}{(1 - R^2)/[n - (k + 1)]} = \dfrac{.899/3}{(1 - .899)/6} = 17.802$

Rejection region: $\alpha = .01$, $\nu_1 = 3$, $\nu_2 = 6$, $F_{.01} = 9.78$
Reject H_0 if $F > 9.78$

Conclusion: Reject H_0. There is sufficient evidence at $\alpha = .01$ to indicate that the model is useful for predicting rate of conversion.

b. H_0: $\beta_1 = 0$
H_a: $\beta_1 \neq 0$

Test statistic: $t = \dfrac{\hat{\beta}_1}{s_{\hat{\beta}_1}} = \dfrac{-.808}{.231} = -3.498$

Rejection region: $\alpha = .05$, df $= n - 4 = 6$, $t_{.025} = 2.447$
Reject H_0 if $t < -2.447$ or $t > 2.447$

Conclusion: Reject H_0. There is sufficient evidence (at $\alpha = .05$) to indicate that atom ratio is a useful predictor of rate of conversion. Since $\hat{\beta}_1$ is a negative value, we can infer that the relationship is negative.

c. A 95% confidence interval for β_2 is $\hat{\beta}_2 \pm t_{.025}s_{\hat{\beta}_2}$ where $t_{.025} = 2.447$ is based on $n - 4 = 6$ df. Substitution yields:

$$-6.38 \pm 2.447(1.93) \Rightarrow -6.38 \pm 4.723 \Rightarrow (-11.103, -1.657)$$

We are 95% confident that the change in the mean rate of conversion is between -11.103 and -1.657 for each additional unit increase in reduction temperature, with atom ratio and acidity support held constant. Note that all values in the interval are negative, indicating a negative relationship between the variables.

4.57 a. $\hat{y} = 22.019 - .181x_1 - .250x_2 - 4.691x_3 + 3.674x_4 + 22.520x_5$

b. $R^2 = 1 - \dfrac{\text{SSE}}{\text{SS(Total)}} = 1 - \dfrac{21886}{54660} = 1 - .4004 = .5996$

Approximately 60% of the sample variation in home-origin trip rate (y) can be explained by the multiple regression model with five independent variables.

c. $s = \sqrt{\text{MSE}} = \sqrt{74.95} = 8.657$

We expect most of the home-origin trip rate (y) values to fall within $2s = 2(8.657) = 17.314$ trips of their least squares predicted values using the multiple regression model.

d. H_0: $\beta_1 = \beta_2 = \beta_3 = \beta_4 = \beta_5 = 0$
H_a: At least one $\beta_i \neq 0$, $i = 1, 2, 3, 4, 5$

Test statistic: $F = 87.45$

Rejection region: $\alpha = .05$, df(Regression) $= 5 =$ numerator df, df(Error) $= 292 =$ denominator df, $F_{.05} \approx 2.21$
Reject H_0 if $F > 2.21$

Conclusion: $\alpha = .05$. There is sufficient evidence to indicate that the model is useful for predicting home-origin trip rate y.

e. H_0: $\beta_1 = 0$
H_a: $\beta_1 < 0$

Test statistic: $t = \dfrac{\hat{\beta}_1}{s_{\hat{\beta}_1}} = \dfrac{-.181}{.039} = -4.64$

Rejection region: $\alpha = .05$, df $=$ df(Error) $= 292$, $t_{.05} \approx 1.645$
Reject H_0 if $t < -1.645$

Conclusion: Reject H_0 at $\alpha = .05$. There is sufficient evidence to indicate that home-origin trip rate (y) and in-vehicle travel time (x_1) are negatively linearly related.

f. A 95% confidence for β_4 is $\hat{\beta}_4 \pm t_{.025} s_{\hat{\beta}_4}$ where $t_{.025} \approx 1.96$ is based on df(Error) $= 292$ df.

$\hat{\beta}_4 = 3.6745$ and $s_{\hat{\beta}_4} = .4027$

Substitution yields:

$$3.6745 \pm (1.96)(.4027) \Rightarrow 3.6745 \pm .7893 \text{ or } (2.8852, 4.4638)$$

Therefore, we are 95% confident that the increase in mean home-origin trip rate y for each additional transit route serving the zone will fall between 2.89 and 4.46 trips, holding the remaining variables constant.

g. $\hat{\beta}_5 \pm t_{.025} s_{\hat{\beta}_5} \Rightarrow 22.52 \pm (1.96)(3.5959) \Rightarrow 22.52 \pm 7.048$ or $(15.472, 29.568)$

Based on the 95% confidence interval, we estimate that the mean home-origin trip rate for zones at the end of a major regional transportation corridor will exceed the mean home-origin trip rate for other zones by at least 15.47 trips and by no more than 29.57 trips.

4.59 a. For a sunny weekday, $x_1 = 0$ and $x_2 = 1$; therefore,

$$\hat{y} = 250 - 700(0) + 100(1) + 5x_3 + 15(0)x_3$$
$$\hat{y} = 350 + 5x_3$$

To graph a straight line, we need a minimum of two points:

x_3	$\hat{y}$
70	700
100	850

For a sunny weekend, $x_1 = 1$ and $x_2 = 1$; therefore,

$$\hat{y} = 250 - 700(1) + 100(1) + 5x_3 + 15(1)x_3$$
$$\hat{y} = -350 + 20x_3$$

Similarly, we need to locate two points which satisfy the equation:

x_3	$\hat{y}$
70	1050
100	1650

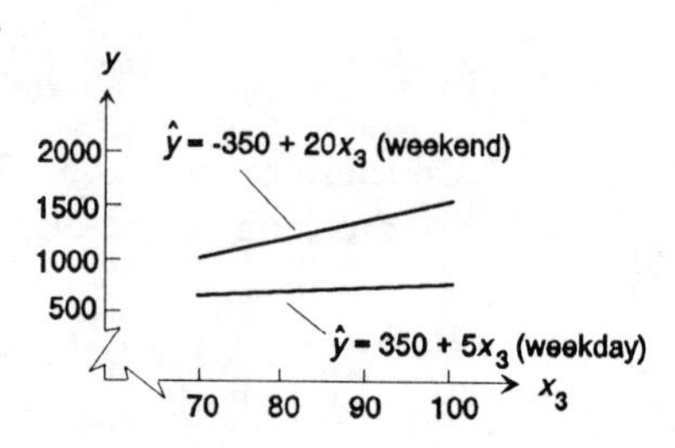

b. H_0: $\beta_4 = 0$
H_a: $\beta_4 \neq 0$

Test statistic: $t = \dfrac{\hat{\beta}_4}{s_{\hat{\beta}_4}} = \dfrac{15}{3} = 5$

Rejection region: $\alpha = .05$, df $= 25$, $t_{.025} = 2.060$
Reject H_0 if $t < -2.060$ or $t > 2.060$

Conclusion: Reject H_0 at $\alpha = .05$. The data indicate that the interaction term is a useful addition to the model.

c. For a sunny weekday with an anticipated high temperature of 95°F, we set $x_1 = 0$, $x_2 = 1$, $x_3 = 95$, and predict:

$$\hat{y} = 250 - 700(0) + 100(1) + 5(95) + 15(0)(95)$$
$$\hat{y} = 825$$

d. The prediction interval in part (d), Exercise 4.59, is smaller than that in Exercise 4.58. The smaller interval indicates a more accurate prediction with the same degree of confidence. Hence, by including the interaction term, we have narrowed the prediction interval considerably. This supports the utility of the interaction term in predicting attendance.

e. Including the interaction term has redefined the coefficients. On a weekday, $x_1 = 0$, and the mean attendance is given by:

$$E(y) = \beta_0 + \beta_1(0) + \beta_2 x_2 + \beta_3 x_3 + \beta_4(0)x_3$$
$$E(y) = \beta_0 + \beta_2 x_2 + \beta_3 x_3$$

On a weekend day, $x_1 = 1$, and the mean attendance is:

$$E(y) = \beta_0 + \beta_1(1) + \beta_2 x_2 + \beta_3 x_3 + \beta_4(1)x_3$$
$$E(y) = \beta_0 + \beta_1 + \beta_2 x_2 + \beta_3 x_3 + \beta_4 x_3$$

Therefore, the difference in mean attendance from weekends to weekdays is:

$$\beta_0 + \beta_1 + \beta_2 x_2 + \beta_3 x_3 + \beta_4 x_3 - (\beta_0 + \beta_2 x_2 + \beta_3 x_3)$$

or

$$\beta_1 + \beta_4 x_3$$

An estimate of the difference in mean attendance from weekends to weekdays is $\hat{\beta}_1 + \hat{\beta}_4 x_3 = -700 + 15x_3$ and not simply -700.

4.61 a. $\hat{y} = 2.14 - .15x_1 + .03x_2 + 2.54x_3 - .34x_4 - .26x_5 - .72x_6$

b. H_0: $\beta_1 = \beta_2 = \cdots = \beta_6 = 0$
H_a: At least one coefficient is nonzero

Test statistic: $F = \dfrac{R^2/k}{(1 - R^2)/[n - (k + 1)]} = \dfrac{.43/6}{.57/201} = 25.27$

Rejection region: $\alpha = .05$, $\nu_1 = 6$, $\nu_2 = 201$, $F_{.05} \approx 2.10$
Reject H_0 if $F > 2.10$

Conclusion: Reject H_0. There is sufficient evidence (at $\alpha = .05$) to indicate that the model is useful for predicting the dissatisfaction level for auto repair services.

c. H_0: $\beta_1 = 0$
H_a: $\beta_1 < 0$

Conclusion: Since $p < .05/2 = .025$, we can reject H_0 at $\alpha = .05$ and conclude that the dissatisfaction will decline as the amount of external information search increases.

d. H_0: $\beta_2 = 0$
H_a: $\beta_2 \neq 0$

Conclusion: Since $p > .05$, we cannot reject H_0 at $\alpha = .05$ and conclude there is not sufficient evidence to indicate that repair cost is a useful predictor of the dissatisfaction level.

e. Since variables x_4, x_5, and x_6 are coded dummy variables concerning service experience, we would test:

H_0: $\beta_4 = \beta_5 = \beta_6 = 0$
H_a: At least one of the coefficient is nonzero

4.63 a. For a city route only, the mean number of riders is:

$$E(y) = \beta_0 + \beta_1 x_1 + \beta_2 + \beta_3 x_1 \text{ since } x_2 = 1$$

For a suburb-city route, $x_2 = 0$; hence, $E(y) = \beta_0 + \beta_1 x_1$.

If the relationship between gasoline price, x_1, and mean number of riders $E(y)$, is the same for the two different bus routes, then $\beta_2 = \beta_3 = 0$. If the relationship between x_1 and $E(y)$ is different for the two routes, then either $\beta_2 \neq 0$ or $\beta_3 \neq 0$, or both. Therefore, we test:

H_0: $\beta_2 = \beta_3 = 0$
H_a: At least one of the coefficients is nonzero

b. For city buses, $\hat{y} = 505 + 40x_1$, whereas, for suburb-city buses, $\hat{y} = 500 + 50x_1$. These relationships are graphed for wholesale gas prices between \$.75 and \$1.50 per gallon:

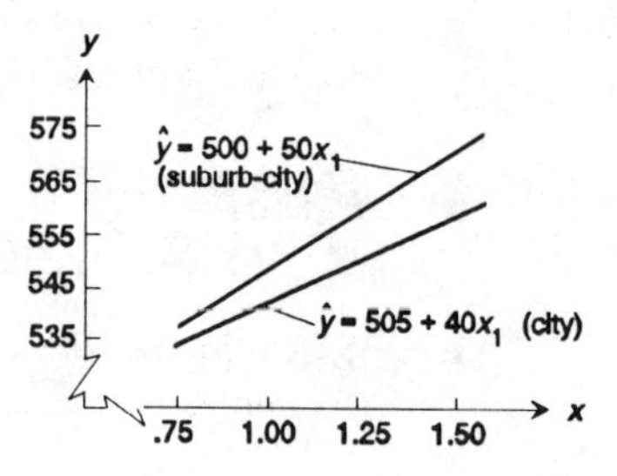

c. H_0: $\beta_3 = 0$
H_a: $\beta_3 \neq 0$

Test statistic: $t = \dfrac{\hat{\beta}_3}{s_{\hat{\beta}_3}} = \dfrac{-10}{3} = -3.33$

Rejection region: $\alpha = .05$, df $= 8$, $t_{.025} = 2.306$
Reject H_0 if $t < -2.306$ or $t > 2.306$

Conclusion: Reject H_0 at $\alpha = .05$. There is sufficient evidence to conclude that the type of bus route and the price of gas interact to affect the number of bus riders. As gas prices increase, suburb-city buses experience a larger share of riders than city buses.

5 Model Building

5.1 a. Since the variable **state of birth** is measured on a nonnumerical scale, it is qualitative. Examples of state of birth are Florida, Ohio, etc.

b. Since the variable **age** is measured using a numerical scale, it is quantitative. Examples of ages are 15 years, 25 years, etc.

c. Since the variable **education level** is measured on a nonnumerical scale, it is qualitative. Examples of education level are high school graduate, college graduate, etc.

d. Since the variable **tenure with firm** is measured using a numerical scale, it is quantitative. We assume that tenure with a firm is the number of years of service. Examples of tenure with a firm are 15 years, 25 years, etc.

e. Since the variable **total compensation** is measured using a numerical scale, it is quantitative. Examples of total compensation are $45,000, $57,843, etc.

f. Since the variable **area of expertise** is measured on a nonnumerical scale, it is qualitative. Examples of areas of expertise are statistics, accounting, etc.

5.3 a. Number of preincident psychological symptoms is a **quantitative** variable ranging from zero to a high value of maybe 10 or 15.

b. Years of experience is a **quantitative** variable ranging from 0 to around 40.

c. Cigarette smoking behavior is a **qualitative** variable. Possible values for this variable is smokes or does not smoke.

d. Level of social support is a **qualitative** variable. Possible values of the variable are low, medium, or high.

e. Marital status is a **qualitative** variable. Possible values of the variable are single, married, divorced, widowed, or separated.

f. Age is a **quantitative** variable with values ranging from a low of 18 to a high of about 65.

g. Ethnic status is a **qualitative** variable. Possible values of this variable are Caucasian, Afro-American, American Indian, Oriental, etc.

h. Exposure to a chemical fire is a **qualitative** variable. Possible values of this variable are exposed or not exposed.

i. Educational level is a **qualitative** variable. Possible values of this variable are below high school, high school diploma, some college, bachelor's degree, master's degree, or doctorate degree.

j. Distance lived from site of incident is a **quantitative** variable, ranging form 0 miles to maybe 40 miles.

k. Gender is a **qualitative** variable with values male or female.

5.5 a. i. First-order
ii. Third-order
iii. First-order
iv. Second-order

b. i. $E(y) = \beta_0 + \beta_1 x$
ii. $E(y) = \beta_0 + \beta_1 x + \beta_2 x^2 + \beta_3 x^3$
iii. $E(y) = \beta_0 + \beta_1 x$
iv. $E(y) = \beta_0 + \beta_1 x + \beta_2 x^2$

c. i. $\beta_1 > 0$ since the slope of the line is positive.
ii. $\beta_3 > 0$ refer to Figure 5.4.
iii. $\beta_1 < 0$ since the slope of the line is negative.
iv. $\beta_2 < 0$ since the parabola opens downward.

5.7 $E(y) = \beta_0 + \beta_1 x + \beta_2 x^2 + \beta_3 x^3$

5.9 The management is advocating a second-order model relating y, assembly time, to x, time since lunch:

$$E(y) = \beta_0 + \beta_1 x + \beta_2 x^2$$

Furthermore, the average assembly time is expected to decrease for small values of x and then gradually increase as x gets larger. Hence, we expect β_2 to be positive and the general shape of the model to look like this:

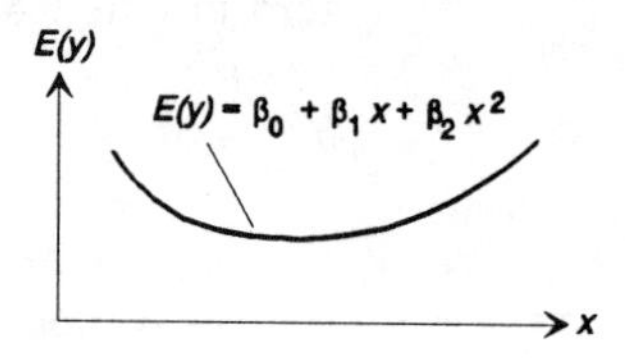

5.11 For $x_1 = 1$:
For $x_2 = 10$: $\hat{y} = 1 + .05(1) + .05(1)(10) = 1.55$
$x_2 = 50$: $\hat{y} = 1 + .05(1) + .05(1)(50) = 3.55$

For $x_1 = 5$:
For $x_2 = 10$: $\hat{y} = 1 + .05(5) + .05(5)(10) = 3.75$
For $x_2 = 50$: $\hat{y} = 1 + .05(5) + .05(5)(50) = 13.75$

For $x = 10$:
For $x_2 = 10$: $\hat{y} = 1 + .05(10) + .05(10)(10) = 6.50$
For $x_2 = 50$: $\hat{y} = 1 + .05(10) + .05(10)(50) = 26.50$

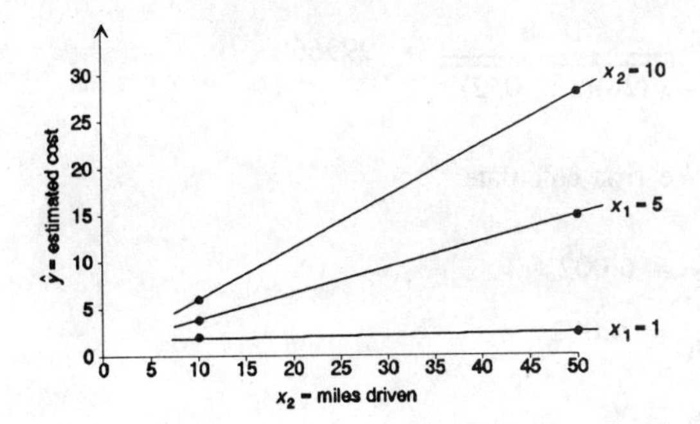

This agrees with our graph in Exercise 5.10, part (c).

5.13 a. The complete second-order model is:

$$E(y) = \beta_0 + \beta_1 x_1 + \beta_2 x_2 + \beta_3 x_1 x_2 + \beta_4 x_1^2 + \beta_5 x_2^2$$

b. $E(y) = \beta_0 + \beta_1 x_1 + \beta_2 x_2$

c. $E(y) = \beta_0 + \beta_1 x_1 + \beta_2 x_2 + \beta_3 x_1 x_2$

d. For fixed x_2, the slope becomes $\beta_1 + \beta_3 x_2$.

e. For fixed x_1, the slope becomes $\beta_2 + \beta_3 x_1$.

5.15 a. We use the following summary information to calculate $\bar{x}$ and s_x for the sample of x-values:

$$n = 7, \sum x^2 = 7651, \sum x = 231$$

Then $\bar{x} = \sum x/n = 231/7 = 33$

$$\text{and } s_x = \sqrt{\frac{\sum x^2 - \frac{(\sum x)^2}{n}}{n - 1}} = \sqrt{\frac{7651 - \frac{(231)^2}{7}}{6}} = \sqrt{4.667} = 2.16$$

Using the coding system for observational data, we have:

$$u = \frac{x - \bar{x}}{s_x} = \frac{x - 33}{2.16}$$

b.

x	30	31	32	33	34	35	36
u	−1.389	−.926	−.436	0	.463	.926	1.389

c. Let $x_1 = x$ and $x_2 = x^2$. We first calculate:

$\sum x_1 = 231$ $\quad\quad$ $\sum x_2 = 7651$

$\sum x_1^2 = 7651$ $\quad\quad$ $SS_{x_1 x_1} = 28$

$\sum x_2^2 = 8{,}484{,}595$ $\quad\quad$ $SS_{x_1 x_2} = 1848$

$\sum x_1 x_2 = 254{,}331$ $\quad\quad$ $SS_{x_2 x_2} = 122{,}052$

Then, $r = \dfrac{SS_{x_1x_2}}{\sqrt{SS_{x_1x_1}SS_{x_2x_2}}} = \dfrac{1848}{\sqrt{(28)(122{,}052)}} = .99966$

d. Let $u_1 = u$ and $u_2 = u^2$. We first calculate:

$\sum u_1 = 0$ $\quad\quad \sum u_2 = 6.002$

$\sum u_1^2 = 6.002$ $\quad\quad SS_{u_1u_1} = 6.002$

$\sum u_2^2 = 9.007$ $\quad\quad SS_{u_1u_2} = 0$

$\sum u_1u_2 = 0$ $\quad\quad SS_{u_2u_2} = 3.860$

Then, $r = \dfrac{SS_{u_1u_2}}{\sqrt{SS_{u_1u_1}SS_{u_2u_2}}} = \dfrac{0}{\sqrt{(6.002)(3.860)}} = 0$

e. Using the SAS multiple regression procedure, we obtain:

$$\hat{y} = 37.5714 - .4629u - 5.3333u^2$$

5.17 a. We use the following summary information to calculate the $\bar{x}$ and s_x for the sample of x values:

$$n = 8, \sum x^2 = 59{,}472.24, \sum x = 680.8$$

Then, $\bar{x} = \dfrac{\sum x}{n} = \dfrac{680.8}{8} = 85.1$

and $s_x = \sqrt{\dfrac{\sum x^2 - \dfrac{(\sum x)^2}{n}}{n - 1}} = \sqrt{\dfrac{59{,}472.24 - \dfrac{(680.8)^2}{8}}{8 - 1}} = 14.814$

Using the coding system for observational data, we have $u = \dfrac{x - \bar{x}}{s_x} = \dfrac{x - 85.1}{14.814}$

b.

x	75.2	91.7	100.3	64.2	81.8	110.2	77.3	80.1
u	−.668	.446	1.026	−1.411	−.223	1.694	−.527	−.338

c. Let $x_1 = x$ and $x_2 = x^2$. We first calculate:

$\sum x_1 = 680.8$ $\quad\quad \sum x_2 = 59{,}472.24$

$\sum x_1^2 = 59{,}472.24$ $\quad\quad SS_{x_1x_1} = 1{,}536.16$

$\sum x_2^2 = 490{,}001{,}910.69$ $\quad\quad SS_{x_2x_2} = 47{,}883{,}494$

$\sum x_1x_2 = 5{,}331{,}419.5$ $\quad\quad SS_{x_1x_2} = 270{,}331.88$

Then, $r = \dfrac{SS_{x_1x_2}}{\sqrt{SS_{x_1x_1}SS_{x_2x_2}}} = \dfrac{270{,}331.88}{\sqrt{1{,}536.16(47.883{,}494)}} = .99675$

d. Let $u_1 = u$ and $u_2 = u^2$. We first calculate:

$\sum u_1 = 0$ $\quad\quad$ $\sum u_2 = 7.000$

$\sum u_1^2 = 7.000$ $\quad\quad$ $SS_{u_1u_1} = 7.000$

$\sum u_2^2 = 13.638$ $\quad\quad$ $SS_{u_2u_2} = 7.513$

$\sum u_1u_2 = 2.7266$ $\quad\quad$ $SS_{u_1u_2} = 2.7266$

$$\text{Then, } r = \frac{SS_{u_1u_2}}{\sqrt{SS_{u_1u_1}SS_{u_2u_2}}} = \frac{2.7266}{\sqrt{7(7.513)}} = .376$$

e. Using MINITAB, the fitted model is $\hat{y} = 110.953 + 14.377u + 7.425u^2$.

5.19 a. β_1 is the difference between the mean job satisfaction of husbands whose wives are employed and the mean job satisfaction of husbands whose wives are unemployed.

From the test, this implies that the mean job satisfaction of husbands whose wives are employed is less than the mean job satisfaction of husbands whose wives are unemployed at $\alpha = .01$.

b. $R^2 = .02$. This implies that only 2% of the sample variation in the husband's job satisfaction scores is explained by the model containing the wife's employment status.

5.21 a. For males $x_5 = 0$ and the model is:

$$E(y) = \beta_0 + \beta_1x_1 + \beta_2x_2 + \beta_3x_3 + \beta_4x_4$$

b. β_1 = difference in mean starting salaries between graduating seniors in Business Administration and Nursing, $\mu_{BA} - \mu_N$

c. β_2 = difference in mean starting salaries between graduating seniors in Engineering and Nursing, $\mu_E - \mu_N$

d. β_3 = difference in mean starting salaries between graduating seniors in Liberal Arts and Sciences and Nursing, $\mu_{LAS} - \mu_N$

e. β_4 = difference in mean starting salaries between graduating seniors in Journalism and Nursing, $\mu_J - \mu_N$

f. For females $x_5 = 1$ and the model is:

$$E(y) = (\beta_0 + \beta_5) + \beta_1x_1 + \beta_2x_2 + \beta_3x_3 + \beta_4x_4$$

g. β_1 = difference in mean starting salaries between males in Business Administration and males in Nursing, $\mu_{BA} - \mu_N$. This is the same as in part (b).

h. β_2 = difference in mean starting salaries between males in Engineering and males in Nursing, $\mu_E - \mu_N$. This is the same as in part (c).

i. β_3 = difference in mean starting salaries between males in Liberal Arts and Sciences and males in Nursing, $\mu_{LAS} - \mu_N$. This is the same as in part (d).

j. β_4 = difference in mean starting salaries between males in Journalism and males in Nursing, $\mu_J - \mu_N$. This is the same as in part (e).

k. For a given college, β_5 is the difference in mean starting salary between females and males, $\mu_F - \mu_M$.

l. To determine if the model is adequate, we test:

H_0: $\beta_1 = \beta_2 = \beta_3 = \beta_4 = \beta_5 = 0$
H_a: At least one $\beta_i \neq 0$ for $i = 1, 2, 3, 4, 5$

Test statistic: $F = 90.022$

The p-value associated with this F value is $p = .0001$. Since this p-value is so small, we reject H_0. There is sufficient evidence to indicate the model is adequate.

The p-values associated with each of the dummy variables are all less than .05. Thus, all are significant. This means that the mean starting salaries for females and males are statistically different. Since the parameter estimate for the gender term is negative, this indicates that the mean starting salary for females is less than the mean starting salary for males. Also, the mean starting salary for nurses is significantly different from the mean starting salaries of all other majors.

The R-square value is .3344. This indicates that 33.44% of the variability in the starting salaries is explained by the model. This is not a particularly high value. There is much variability in starting salaries that is not explained by this model.

5.23 a. Consider the model:

$$E(y) = \beta_0 + \beta_1 x_1 + \beta_2 x_2 + \beta_3 x_3$$

where

$$x_1 = \begin{cases} 1 & \text{if market concentration is moderate} \\ 0 & \text{if not} \end{cases}$$

$$x_2 = \begin{cases} 1 & \text{if market concentration is high} \\ 0 & \text{if not} \end{cases}$$

$$x_3 = \begin{cases} 1 & \text{if consumer industry type} \\ 0 & \text{if not} \end{cases}$$

b. Let μ_{ij} be the mean excess capacity for market concentration $i(i = 1$ (low), 2 (moderate), 3(high)) and industry type $j(j = 1$ (producer), 2 (consumer)). Then,

$\beta_0 = \mu_{11}$ (mean excess capacity for market concentration and producer type industry)

$\beta_1 = \mu_{2j} - \mu_{1j}$ (difference in mean weekly sales between moderate and low market concentration for any level of industry type, $j = 1, 2$)

$\beta_2 = \mu_{3j} - \mu_{1j}$ (difference in mean weekly sales between consumer and low market concentration for any level of industry type, $j = 1, 2$)

$\beta_3 = \mu_{i2} - \mu_{i1}$ (difference in mean weekly sales between consumer and producer industry types for any level of market concentration, $i = 1, 2, 3$)

c. $E(y) = \beta_0 + \beta_1 x_1 + \beta_2 x_2 + \beta_3 x_3 + \beta_4 x_1 x_3 + \beta_5 x_1 x_3$ where x_1, x_2, and x_3 are defined as in (a).

d. Using the same notation as in (b), then

$\beta_0 = \mu_{11}$

$\beta_1 = \mu_{21} - \mu_{11}$ (difference in mean weekly sales between moderate and low market concentration for producer type industries)

$\beta_2 = \mu_{31} - \mu_{11}$ (difference in mean weekly sales between high and low market concentration for producer type industries)

$\beta_3 = \mu_{12} - \mu_{11}$ (difference in mean weekly sales between consumer and producer industries for low market concentration)

$\beta_4 = (\mu_{22} - \mu_{12}) - (\mu_{21} - \mu_{11})$
(difference between the difference in mean weekly sales for moderate and low market concentrations for consumer type industries and producer type industries)

$\beta_5 = (\mu_{32} - \mu_{12}) - (\mu_{31} - \mu_{11})$
(difference between the difference in mean weekly sales for high and low market concentrations for consumer type industries and producer type industries)

e. Test H_0: $\beta_4 = \beta_5 = 0$ using a partial F test.

5.25 a. Let x_2 be the quantitative variable. The complete second-order model is:

$$E(y) = \beta_0 + \beta_1 x_2 + \beta_2 x_2^2$$

b. Let $x_3 = \begin{cases} 1 & \text{if benzene} \\ 0 & \text{if not} \end{cases}$ $x_4 = \begin{cases} 1 & \text{if toluene} \\ 0 & \text{if not} \end{cases}$

$x_5 = \begin{cases} 1 & \text{if chloroform} \\ 0 & \text{if not} \end{cases}$ $x_6 = \begin{cases} 1 & \text{if methanol} \\ 0 & \text{if not} \end{cases}$

The new model would be:

$$E(y) = \beta_0 + \beta_1 x_2 + \beta_2 x_2^2 + \beta_3 x_3 + \beta_4 x_4 + \beta_5 x_5 + \beta_6 x_6$$

c. The new complete second-order model is:

$$E(y) = \beta_0 + \beta_1 x_2 + \beta_2 x_2^2 + \beta_3 x_3 + \beta_4 x_4 + \beta_5 x_5 + \beta_6 x_6 + \beta_7 x_2 x_3 + \beta_8 x_2 x_4 + \beta_9 x_2 x_5 + \beta_{10} x_2 x_6 + \beta_{11} x_2^2 x_3 + \beta_{12} x_2^2 x_4 + \beta_{13} x_2^2 x_5 + \beta_{14} x_2^2 x_6$$

d. The response curves will have the same shape but different y-intercepts only if all of the interaction terms are 0. Therefore, if $\beta_7 = \beta_8 = \cdots = \beta_{14} = 0$

e. The response curves will be parallel lines if all interaction and quadratic variables are 0. Therefore, if $\beta_2 = \beta_7 = \beta_8 = \cdots = \beta_{14} = 0$

f. The response curves will all be identical if all interaction and qualitative variables are 0. Therefore, if $\beta_3 = \beta_4 = \cdots = \beta_{14} = 0$

5.27 a. First, we test to see if the overall model is adequate for predicting mass burning rate:

H_0: $\beta_1 = \beta_2 = \beta_3 = 0$
H_a: At least one of the coefficients is not 0

Test statistic: From the printout, the test statistic is $F = 14.42$.

The p-value is .001. Since the p-value is so small, we reject H_0 for any value of $\alpha > .001$. Using $\alpha = .05$, we reject H_0. There is sufficient evidence to indicate that the overall model is adequate for predicting mass burning rate.

$R^2 = .812$. This implies that 81.2% of the sample variation in the mass burning rates is explained by the model containing brake power and type of fuel.

The estimate of the standard deviation is $s = 8.057$. This means that almost all of the observed values of mass burning rate will fall within $2s$ or 2(8.057) or 16.114 units of their respective least squares predicted values.

b. To determine if brake power and fuel type interact, we test:

H_0: $\beta_4 = \beta_5 = 0$
H_a: At least one β_4 or β_5 is different from 0

Test statistic: $F = \dfrac{(\text{SSE}_R - \text{SSE}_C)/(k - g)}{\text{SSE}_C/[n - (k + 1)]} = \dfrac{(649.09 - 203.01)/(5 - 3)}{203.01/[14 - (5 + 1)]} = 8.79$

Rejection region: $\alpha = .01$, $\nu_1 = k - g = 2$, $\nu_2 = n - (k + 1) = 8$, $F_{.01} = 8.65$
Reject H_0 if $F > 8.65$

Conclusion: Reject H_0 at $\alpha = .01$. There is sufficient evidence to indicate that brake power and fuel type interact to affect mass burning rate.

c. For fuel type **advanced timing:** $x_2 = 0$ and $x_3 = 0$

$$E(y) = \beta_0 + \beta_1x_1 + \beta_2(0) + \beta_3(0) + \beta_4x_1(0) + \beta_5x_1(0)$$
$$= \beta_0 + \beta_1x_1$$

The slope of the $y - x_1$ line is β_1. The estimate of the slope is $\hat{\beta}_1 = 7.815$.

For fuel type **DF-2**: $x_2 = 1$ and $x_3 = 0$

$$E(y) = \beta_0 + \beta_1x_1 + \beta_2(1) + \beta_3(0) + \beta_4x_1(1) + \beta_5x_1(0)$$
$$= (\beta_0 + \beta_2) + (\beta_1 + \beta_4)x_1$$

The slope of the $y - x_1$ line is $\beta_1 + \beta_4$. The estimate of the slope is $\hat{\beta}_1 + \hat{\beta}_4 = 7.815 - 5.675 = 2.14$.

For fuel type **Blended:** $x_2 = 0$ and $x_3 = 1$

$$E(y) = \beta_0 + \beta_1x_1 + \beta_2(0) + \beta_3(1) + \beta_4x_1(0) + \beta_5x_1(1)$$
$$= (\beta_0 + \beta_3) + (\beta_1 + \beta_5)x_1$$

The slope of the $y - x_1$ line is $\beta_1 + \beta_5$. The estimate of the slope is $\hat{\beta}_1 + \hat{\beta}_5 = 7.815 - 2.950 = 4.865$.

5.29 a. The first order model would be:

$$E(y) = \beta_0 + \beta_1x_1 + \beta_2x_2 + \beta_3x_3$$

where x_1 and x_2 are quantitative and x_3 is a qualitative variable.

b. The rate of change in log odds ratio based on craft occupation inclusion is given by β_3. Then, to test if the mean log odds ratio is smaller for craft occupations than for noncraft occupations, test H_0: $\beta_3 = 0$ versus H_a: $\beta_3 < 0$ using a t test.

c. The complete second-order model is:

$$E(y) = \beta_0 + \beta_1 x_1 + \beta_2 x_2 + \beta_3 x_1 x_2 + \beta_4 x_1^2 + \beta_5 x_2^2 + \beta_6 x_3 + \beta_7 x_1 x_3 + \beta_8 x_2 x_3 + \beta_9 x_1 x_2 x_3 + \beta_{10} x_1^2 x_3 + \beta_{11} x_2^2 x_3$$

d. We wish to see if the coefficients of x_1 are zero, so test:

H_0: $\beta_1 = \beta_3 = \beta_4 = \beta_7 = \beta_9 = \beta_{10} = 0$ using a partial F test.

e. A possible set of contour lines might look like:

where $x_{2,0}$, $x_{2,1}$, $x_{2,2}$ are specific values of x_2

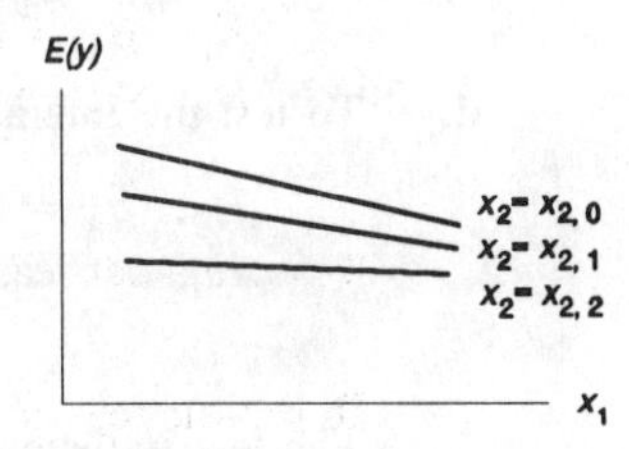

5.31 a. Age represents a **quantitative** variable ranging from a low value of perhaps 16 to a high of 70.

b. Years in therapy can range between say 0 and 80. This is a **quantitative** independent variable.

c. Highest educational degree would probably be broken down into five levels: none, high school diploma, bachelor's degree, master's degree, or doctorate. These categories are indicative of a **qualitative** variable.

d. Job classification is categorical in nature and therefore a **qualitative** independent variable. Some possible "levels" are laborer, foreman, secretary, administrative assistant, manager, vice president, etc.

e. It is impossible to assign numerical values to classifications such as: single, married, divorced, widowed, separated, other. Thus, marital status is a **qualitative** variable.

f. Religious preference is also **qualitative**. Four possible levels are: Protestant, Catholic, Jewish, other.

g. IQ is considered a **quantitative** variable with a wide range of values from 0 to 175 or more.

h. Sex is a **qualitative** variable with the two levels male and female.

5.33 a. $E(y) = \beta_0 + \beta_1 x_1 + \beta_2 x_2 + \beta_3 x_3 + \beta_4 x_1 x_2 + \beta_5 x_1 x_3$

where y = finish time $\qquad x_1$ = number of jobs

$$x_2 = \begin{cases} 1 & \text{if A} \\ 0 & \text{if not} \end{cases} \qquad x_3 = \begin{cases} 1 & \text{if B} \\ 0 & \text{if not} \end{cases}$$

b. To test whether the model is useful, we test:

H_0: $\beta_1 = \beta_2 = \beta_3 = \beta_4 = \beta_5 = 0$
H_a: At least one $\beta_i \neq 0$

The test statistic is $F = 4.895$.

The p-value $= .0394$.

For $\alpha = .05$, we reject H_0 (since $\alpha > p$-value) and conclude the model is useful in predicting finish time.

c. $E(y) = \beta_0 + \beta_1 x_1 + \beta_2 x_2 + \beta_3 x_3$

d. To test the interaction terms, we test:

H_0: $\beta_4 = \beta_5 = 0$
H_a: At least one $\beta_i \neq 0$

The test statistic is $F = \dfrac{(\text{SSE}_R - \text{SSE}_C)/(k - g)}{\text{SSE}_C/[n - (k + 1)]} = \dfrac{(38.289 - 21.443)/(5 - 3)}{21.443/[12 - (5 + 1)]} = 2.36$

Rejection region: $\alpha = .05$, $\nu_1 = 2$, $\nu_2 = 6$, $F_{.05} = 5.14$
Reject H_0 if $F > 5.14$

Conclusion: Do not reject H_0. There is insufficient evidence to indicate the interaction terms are useful at $\alpha = .05$.

5.35 a. The two independent variables, packaging and location, are qualitative. The two types of packaging generate one term in the model, while the four locations induce three terms. A main effect model is:

$$E(y) = \beta_0 + \beta_1 x_1 + \beta_2 x_2 + \beta_3 x_3 + \beta_4 x_4$$

where $x_1 = \begin{cases} 1 & \text{if } P_2 \\ 0 & \text{otherwise} \end{cases}$ $\quad x_3 = \begin{cases} 1 & \text{if } L_3 \\ 0 & \text{otherwise} \end{cases}$

$x_2 = \begin{cases} 1 & \text{if } L_2 \\ 0 & \text{otherwise} \end{cases}$ $\quad x_4 = \begin{cases} 1 & \text{if } L_4 \\ 0 & \text{otherwise} \end{cases}$

This model assumes that Packaging and Location affect total sales independently.

b. If the previous assumption is incorrect and Packaging and Location do interact, then a more appropriate model is:

$$E(y) = \beta_0 + \beta_1 x_1 + \beta_2 x_2 + \beta_3 x_3 + \beta_4 x_4 + \beta_5 x_1 x_2 + \beta_6 x_1 x_3 + \beta_7 x_1 x_4$$

There are eight parameters in this model. There are also a total of eight Packaging-Location combinations, which permits an estimate of $E(y)$ for each combination.

c. H_0: $\beta_5 = \beta_6 = \beta_7 = 0$
H_a: At least one of β_5 or β_6 or β_7 is different from 0

Test statistic: $F = \dfrac{(\text{SSE}_\text{R} - \text{SSE}_\text{C})/(k - g)}{\text{SSE}_\text{C}/[n - (k + 1)]} = \dfrac{(422.36 - 346.65)/(7 - 4)}{346.65/[40 - (7 + 1)]} = 2.33$

Rejection region: $\alpha = .05$, $\nu_1 = k - g = 3$, $\nu_2 = n - (k + 1) = 32$, $F_{.05} \approx 2.92$
Reject H_0 if $F > 2.92$

Conclusion: Do not reject H_0 at $\alpha = .05$. There is insufficient evidence to indicate that the interaction between location and packaging is important in estimating mean weekly sales.

This implies that the company can choose the location and packaging type independently of each other. The choice of packaging type does not depend on location.

5.37 a. The main effects model is:

$$E(y) = \beta_0 + \beta_1 x_1 + \beta_2 x_2$$

where x_1 = employee's years of education $\quad x_2 = \begin{cases} 1 & \text{if employee is certified} \\ 0 & \text{if not} \end{cases}$

b. To include "sheepskin screening," we need to include an interaction term:

$$E(y) = \beta_0 + \beta_1 x_1 + \beta_2 x_2 + \beta_3 x_1 x_2$$

c. The complete second-order model is:

$$E(y) = \beta_0 + \beta_1 x_1 + \beta_2 x_1^2 + \beta_3 x_2 + \beta_4 x_1 x_2 + \beta_5 x_1^2 x_2$$

5.39 a. Since the model includes both interaction and curvature, we envision two response curves, one for each level of the qualitative variable x_2, mode of transportation (rail or truck). The curves have different shapes due to the interaction between distance x_1 and mode x_2.

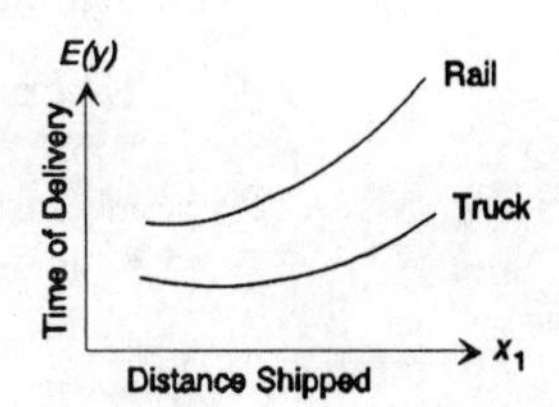

b. The coefficients of the terms involving x_1^2 are β_2 and β_5. Thus, to determine if curvature is present in the relationship between the mean delivery time and distance, we would test:

H_0: $\beta_2 = \beta_5 = 0$
H_a: At least one of the coefficients of β_2 and β_5 is nonzero

c. For a fixed distance, x_1, the mean delivery time by rail is:

$$E(y) = (\beta_0 + \beta_3) + (\beta_1 + \beta_4)x_1 + (\beta_2 + \beta_5)x_1^2$$

The mean delivery time by truck is:

$$E(y) = \beta_0 + \beta_1 x_1 + \beta_2 x_1^2$$

If the mean delivery times are to be identical, then β_3 and β_4, and β_5 must all equal zero. Therefore, we would test:

H_0: $\beta_3 = \beta_4 = \beta_5 = 0$
H_a: At least one of the coefficients, β_3, β_4, and β_5, is nonzero

5.41 a. Lead-free gasoline at the full-service pumps implies $x_1 = 0, x_2 = 1$, and $x_3 = 1$. The mean sales are:

$$E(y) = \beta_0 + \beta_1(0) + \beta_2(1) + \beta_3(1) + \beta_4(0)(1) + \beta_5(1)(1)$$

or

$$E(y) = \beta_0 + \beta_2 + \beta_3 + \beta_5$$

Therefore, an estimate of the mean sales is:

$$\hat{y} = \hat{\beta}_0 + \hat{\beta}_2 + \hat{\beta}_3 + \hat{\beta}_5 = 4 - 1 - 1 + 3 = 5$$

b. The estimated mean sales of regular gasoline at full-service pumps ($x_1 = 0, x_2 = 0, x_3 = 1$) is:

$$\hat{y} = 4 - 2(0) - 0 - 1 + 2(0)(1) + 3(0)(1) = 3$$

The estimated mean sales of regular gasoline at self-service pumps ($x_1 = 0, x_2 = 0, x_3 = 0$) is:

$$\hat{y} = 4 - 2(0) - 0 - 0 + 2(0)(0) + 3(0)(0) = 4$$

The estimated difference in mean sales is $3 - 4 = -1$.

c. An estimate of premium full-service mean sales ($x_1 = 1, x_2 = 0, x_3 = 1$) is $\hat{y} = 4 - 2 - 1 + 2 = 3$. An estimate of premium self-service mean sales ($x_1 = 1, x_2 = 0, x_3 = 0$) is $\hat{y} = 4 - 2 = 2$. This estimated difference between the mean sales for premium full-service and premium self-service is 1, as compared to a difference of -1 from part (b). If we tabulate and graph the estimated mean sales for each possible combination of types of service, we may detect a more revealing indication of interaction:

Estimated Mean Sales

Type of Gasoline	Type of Service Full	Self
Regular	3	4
Premium	3	2
Lead-free	5	3

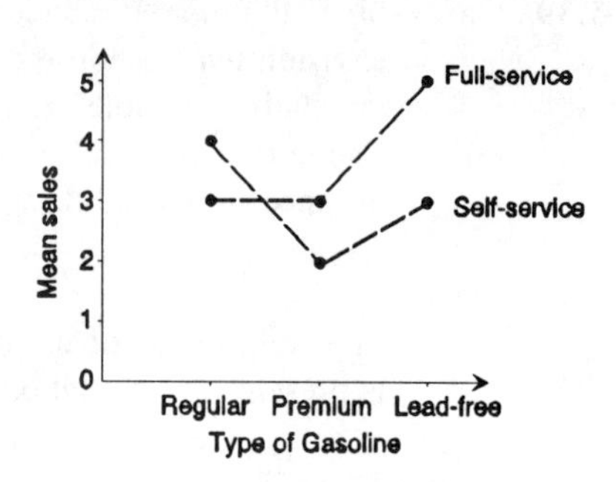

Comparing the graph to the one we constructed in Exercise 5.40, part (d), we find similarities in the mean sales for regular and premium gasoline. However, our intuition about lead-free sales appears to be misguided.

5.43 a. The complete second-order model relating college GPA (y) to high school GPA (x_1) and disability (x_2) is:

$$E(y) = \beta_0 + \beta_1 x_1 + \beta_2 x_1^2 + \beta_3 x_2 + \beta_4 x_1 x_2 + \beta_5 x_1^2 x_2$$

The sketch of a possible relationship is:

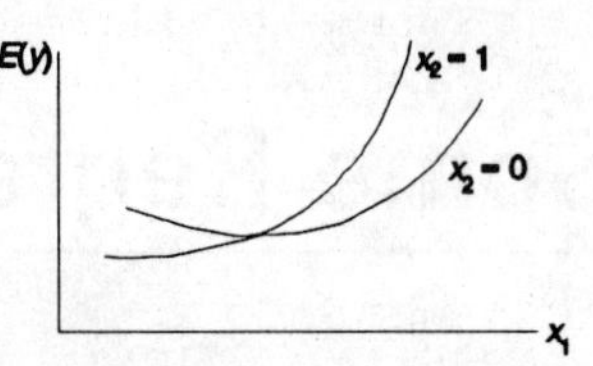

b. The complete second-order model relating college GPA (y) to Internal-External Scale (x_3) and Counseling Relationship Inventory (x_4) is:

$$E(y) = \beta_0 + \beta_1 x_3 + \beta_2 x_4 + \beta_3 x_3 x_4 + \beta_4 x_3^2 + \beta_5 x_4^2$$

c. The first-order model relating college GPA to all four independent variables is:

$$E(y) = \beta_0 + \beta_1 x_1 + \beta_2 x_2 + \beta_3 x_3 + \beta_4 x_4$$

d. The first-order model relating college GPA to all four independent variables is where the relationship between y and the quantitative variables is affected by the disability:

$$E(y) = \beta_0 + \beta_1 x_1 + \beta_2 x_2 + \beta_3 x_3 + \beta_4 x_4 + \beta_5 x_1 x_2 + \beta_6 x_2 x_3 + \beta_7 x_2 x_4$$

6 Some Regression Pitfalls

6.1 If the departures from the assumptions are not too great, the multiple regression analysis will still yield results with (for all practical purposes) the reliability stated in Chapter 4. Flagrant deviations from the assumptions, though, can create problems. The assumption that $E(\epsilon) = 0$ deals with model selection and whether or not additional independent variables need to be included in the model. The remaining assumptions deal with the distribution of the estimators. If the assumptions are violated, any inferences made regarding these estimators may not be sound.

6.3 Multicollinearity first increases the likelihood of rounding errors in our calculations. Second, the results themselves may not seem to make sense (i.e., a significant F test for model utility but no significant t tests among the individual parameter estimates). A third problem is that a high correlation between variables may affect the sign of the parameter estimates, a positive slope is obtained when a negative relationship was expected.

6.5 When multicollinearity is present, one way of dealing with the problem is to remove all but one of the highly correlated variables, possibly by using stepwise regression. When using the model for estimation and prediction only, it is not necessary to drop any independent variables. It is necessary, though, to make sure the values of the x variables fall within the ranges of x used in the experiment. The presence of multicollinearity when making inferences about the β's in the model is a dangerous problem. The solution is to use a designed experiment to break up the patterns of multicollinearity. To reduce rounding errors in polynomial models, it is useful to code the x values so that the correlation between the subsequent powers of x is reduced. Another way of reducing rounding errors due to multicollinearity is to use ridge regression which provides biased estimates of the β's in the model, but these estimates have smaller standard errors than their least squares analogs.

6.7 Since the correlation between Importance and Support is fairly high (.6991), we should drop one of the two variables from the regression analysis. Since the correlation between Support and Replace (−.0531) is smaller than the correlation between Importance and Replace (.2682), we would recommend deleting the variable Importance and keeping the variables Support and Replace.

6.9 When fitting the pth-order polynomial regression model, two requirements must be met:

1. The number of levels of x must be greater than or equal to $(p + 1)$. For a second-order model $(p = 2)$, we must have at least $p + 1 = 3$ levels of x.
2. The sample size n must be greater than $(p + 1)$ in order to allow sufficient degrees of freedom for estimating σ^2. For a second-order model, we must have n greater than $(p + 1) = 3$.

Note that for our sample data, requirement #1 is satisfied since $x = 1, 2$, or 5. However, requirement #2 is not satisfied since $n = 3$. If we attempt to fit this model, we will have $n - (k + 1) = n - 3 = 0$ df for estimating σ^2 and, thus, will be unable to test model adequacy.

6.11 a. Using SAS to fit the model, we get:

```
Model: MODEL1
Dependent Variable: Y
                        Analysis of Variance
                       Sum of           Mean
Source     DF         Squares         Square      F Value      Prob>F
Model       1       494.28131      494.28131      253.370      0.0001
Error      23        44.86909        1.95083
C Total    24       539.15040

     Root MSE       1.39672      R-square       0.9168
     Dep Mean      12.52800      Adj R-sq       0.9132
     C.V.          11.14880

                        Parameter Estimates

                 Parameter     Standard      T for H0:
Variable  DF      Estimate        Error    Parameter=0  Prob > |T|

INTERCEP   1      2.743278   0.67520594          4.063      0.0005
X1         1      0.800976   0.05032017         15.918      0.0001
```

The least squares equation is $\hat{y} = 2.74 + .801x_1$.

The p-value associated with the tar content (x_1) variable is $p = .0001$. For $\alpha > .0001$, tar content is a useful predictor of carbon monoxide content.

b. Using SAS to fit the model, we get:

```
Model: MODEL1
Dependent Variable: Y
                       Analysis of Variance
                        Sum of           Mean
Source      DF         Squares         Square    F Value         Prob>F
Model        1       462.25591      462.25591    138.266         0.0001
Error       23        76.89449        3.34324
C Total     24       539.15040

    Root MSE        1.82845       R-square          0.8574
    Dep Mean       12.52800       Adj R-sq          0.8512
    C.V.           14.59493
                       Parameter Estimates

                Parameter        Standard     T for H0:
Variable  DF     Estimate           Error   Parameter=0  Prob > |T|

INTERCEP   1     1.664666      0.99360176         1.675      0.1074
X2         1    12.395406      1.05415188        11.759      0.0001
```

The least squares equation is $\hat{y} = 1.66 + 12.395x_2$.

The p-value associated with the nicotine content (x_2) variable is $p = .0001$. For $\alpha > .0001$, nicotine content is a useful predictor of carbon monoxide content.

c. Using SAS to fit the model, we get:

```
Model: MODEL1
Dependent Variable: Y
                        Analysis of Variance
                        Sum of          Mean
Source       DF        Squares        Square      F Value    Prob>F
Model         1      116.05651     116.05651        6.309    0.0195
Error        23      423.09389      18.39539
C Total      24      539.15040
     Root MSE        4.28898      R-square         0.2153
     Dep Mean       12.52800      Adj R-sq         0.1811
     C.V.           34.23519
                        Parameter Estimates
                    Parameter       Standard     T for H0:
Variable  DF         Estimate          Error   Parameter=0   Prob >|T|
INTERCEP   1       -11.795271     9.72162623        -1.213      0.2373
X3         1        25.068198     9.98028209         2.512      0.0195
```

The least squares equation is $\hat{y} = -11.795 + 25.068x_3$.

The p-value associated with the weight (x_3) variable is $p = .0195$. For $\alpha > .0195$, weight is a useful predictor of carbon monoxide content.

d. We see the estimate signs have changed dramatically when compared to the model with all three variables present.

6.13 There are two signs that multicollinearity may be a problem in this exercise. First, the F test is very significant, while the individual t tests are not as significant. Second, the AVGSHIP variable is expected to have a positive relationship with LABOR. The parameter estimate is negative. Both of these signs indicate multicollinearity is present in the model.

6.15 Two levels of x_1 and two levels of x_2 are needed to fit the model because there is only a linear relationship between y and each of the independent variables.

Since $n - (k + 1) = n - (3 + 1) = n - 4$ must be positive in order to estimate σ^2, a sample size of at least 5 is needed.

6.17 a. To test for model adequacy, we test:

H_0: $\beta_1 = \beta_2 = \beta_3 = \beta_4 = 0$
H_a: At least one $\beta_i \neq 0$, $i = 1, 2, 3, 4$

The test statistic is $F = 37.204$.

The p-value is .0007.

For any $\alpha > .0007$, we can reject H_0. There is sufficient evidence to indicate the model is useful for any $\alpha > .0007$.

b. Because the coefficient associated with chest depth has a different sign than what is expected, chest depth is probably correlated with some of the other independent variables.

c. This small value of t is probably caused by a high correlation between weight and some of the other independent variables.

d. After calculating the correlations, we find:

$$r_{12} = .327,\ r_{13} = .231,\ r_{14} = .166,\ r_{23} = .790,\ r_{24} = .791,\ r_{34} = .881$$

The high values of some of these correlations (.790, .791, and .881) indicate multicollinearity might be present.

6.19 a. Some preliminary calculations are:

$$\sum x_1 = 450.1 \qquad \sum y = 1800 \qquad \sum x_1 y = 54{,}013.5$$

$$\sum x_1^2 = 13{,}906.27 \qquad \sum y^2 = 216{,}900.86$$

$$SS_{x_1 y} = \sum x_1 y - \frac{\sum x_1 \sum y}{n} = 54{,}013.5 - \frac{450.1(1800)}{15} = 1.5$$

$$SS_{x_1 x_1} = \sum x_1^2 - \frac{\left(\sum x_1\right)^2}{n} = 13{,}906.27 - \frac{450.1^2}{15} = 400.26933$$

$$SS_{yy} = \sum y^2 - \frac{\left(\sum y\right)^2}{n} = 216{,}900.86 - \frac{1800^2}{15} = 900.86$$

The correlation coefficient is:

$$r = \frac{SS_{x_1 y}}{\sqrt{SS_{x_1 x_1} SS_{yy}}} = \frac{1.5}{\sqrt{400.26933(900.86)}} = .0025$$

Because the value of r is so small, there is no evidence of a linear relationship between x_1 and y.

b. Some preliminary calculations are:

$$\sum x_2 = 1049.8 \qquad \sum y = 1800 \qquad \sum x_2 y = 126{,}844.85$$

$$\sum x_2^2 = 77{,}911.98 \qquad \sum y^2 = 216{,}900.86$$

$$SS_{x_2 y} = \sum x_2 y - \frac{\sum x_2 \sum y}{n} = 126{,}844.85 - \frac{1049.8(1800)}{15} = 868.86$$

$$SS_{x_2 x_2} = \sum x_2^2 - \frac{\left(\sum x_2\right)^2}{n} = 77{,}911.98 - \frac{1049.8^2}{15} = 4439.97733$$

$$SS_{yy} = \sum y^2 - \frac{\left(\sum y\right)^2}{n} = 216{,}900.86 - \frac{1800^2}{15} = 900.86$$

The correlation coefficient is:

$$r = \frac{SS_{x_2 y}}{\sqrt{SS_{x_2 x_2} SS_{yy}}} = \frac{868.86}{\sqrt{4439.97733(900.86)}} = .434$$

Since the value of r is not very close to 1, there is little evidence of a linear relationship between x_2 and y.

c. Based on the values of the correlation coefficients computed in parts (a) and (b), there is no evidence that the model will be useful for predicting the sale price.

d. Using SAS, the output is:

```
DEP VARIABLE: Y
                              ANALYSIS OF VARIANCE

SOURCE          DF    SUM OF SQUARES    MEAN SQUARE       F VALUE      PROB > F
MODEL            2         900.72221      450.36111     39222.343        0.0001
ERROR           12        0.13778711     0.01148226
C TOTAL         14         900.86000

          ROOT MSE         0.1071553       R-SQUARE        0.9998
          DEP MEAN               120       ADJ R-SQ        0.9998
              C.V.        0.08929609

                            PARAMETER ESTIMATES

                        PARAMETER ESTIMATE        STANDARD           T FOR HO:
VARIABLE          DF                                 ERROR         PARAMETER=0

INTERCEP           1          -45.15413641      0.61141807             -73.851
X1                 1            3.09700789      0.01227443             252.314
X2                 1            1.03185903     0.003684173             280.079

VARIABLE          DF            PROB > |T|

INTERCEP           1                0.0001
X1                 1                0.0001
X2                 1                0.0001
```

From the printout, the least squares line is $\hat{y} = -45.154 + 3.097x_1 + 1.032x_2$.

To determine if the model is adequate, we test:

H_0: $\beta_1 = \beta_2 = 0$
H_a: At least one $\beta_i \neq 0$

The test statistic is $F = 39{,}222.343$

The p-value is .0001. Since this p-value is so small, H_0 is rejected. There is sufficient evidence to indicate the model is adequate for predicting sale price.

The value of R^2 is .9998. This indicates that 99.98% of the sample variation in the sale prices is explained by the model including x_1 and x_2.

These results disagree with those in part (c).

e. Some preliminary calculations are:

$\sum x_1 = 450.1$ $\sum x_1^2 = 13,906.27$ $\sum x_1 x_2 = 30,304.09$

$\sum x_2 = 1049.8$ $\sum x_2^2 = 77,911.98$

$$SS_{x_1 x_2} = \sum x_1 x_2 - \frac{\sum x_1 \sum x_2}{n} = 30,304.09 - \frac{450.1(1049.8)}{15} = -1196.90867$$

$$SS_{x_1 x_1} = \sum x_1^2 - \frac{\left(\sum x_1\right)^2}{n} = 13,906.27 - \frac{450.1^2}{15} = 400.26933$$

$$SS_{x_2 x_2} = \sum x_2^2 - \frac{\left(\sum x_2\right)^2}{n} = 77,911.98 - \frac{1049.8^2}{15} = 4439.97733$$

The correlation coefficient is:

$$r = \frac{SS_{x_1 x_2}}{\sqrt{SS_{x_1 x_1} SS_{x_2 x_2}}} = \frac{-1196.90867}{\sqrt{400.26933(4439.97733)}} = -.8978$$

Because the correlation coefficient between x_1 and x_2 is very close to -1, it implies that x_1 and x_2 are highly correlated.

f. In this case, we would not want to throw out a redundant variable. The models with just one independent variable are not significant. However, the model with both independent variables, even though they are highly correlated, is very significant.

6.21 The researcher's sample size will not permit him to estimate σ^2. Thus, no estimate of the variation in the y values exists and no inferences can be made.

7 Residual Analysis

7.1 a. The predicted values are obtained by substituting the observed values of x into the least squares prediction equation:

$$\hat{y} = 2.588 + .541x$$

The regression residuals are then computed as shown in the table.

	Observed Value	Predicted Value	Residual
x	y	$\hat{y}$	$(y - \hat{y})$
−2	1.1	1.506	−.406
−2	1.3	1.506	−.206
−1	2.0	2.047	−.047
−1	2.1	2.047	.053
0	2.7	2.588	.112
0	2.8	2.588	.212
1	3.4	3.129	.271
1	3.6	3.129	.471
2	4.0	3.670	.330
2	3.9	3.670	.230
3	3.8	4.211	−.411
3	3.6	4.211	−.611

b. The plot of the residuals versus x is shown at right.

Notice that the plot reveals a quadratic upside down (U-shaped) trend. For small and large values of x, the residuals are negative, while for intermediate values of x, the residuals are positive. This pattern indicates a lack of fit in the straight-line model and suggests that a term to introduce curvature in the model is necessary. To verify this, we would fit the quadratic model:

$$E(y) = \beta_0 + \beta_1 x + \beta_2 x^2$$

and test for curvature, i.e., test H_0: $\beta_2 = 0$

7.3 a. Preliminary calculations are shown below:

$n = 14$ $\qquad$ $SS_{xx} = 56$

$$\hat{\beta}_1 = \frac{SS_{xy}}{SS_{xx}} = \frac{-11.6}{56} = -.207142857 \approx -.207$$

$\sum x = 462$ $\qquad$ $SS_{xy} = -11.6$

$\sum y = 469.2$

$\sum x^2 = 15{,}302$ $\qquad\qquad \bar{x} = 33$

$\sum y^2 = 15{,}910.74$ $\qquad\qquad \bar{y} = 33.51428571$

$\sum xy = 15{,}472$

$$\hat{\beta}_0 = \bar{y} - \hat{\beta}_1\bar{x} = \frac{469.2}{14} - (-.207142857)\frac{462}{14} = 40.35$$

The least squares prediction equation is $\hat{y} = 40.35 - .207x$.

b. The least squares equation above is used to obtain the predicted values and residuals shown below:

x	Observed Value y	Predicted Value $\hat{y}$	Residual $(y - \hat{y})$
30	29.5	34.140	−4.640
30	30.2	34.140	−3.940
31	32.1	33.933	−1.833
31	34.5	33.933	0.567
32	36.3	33.726	2.574
32	35.0	33.726	1.274
33	38.2	33.519	4.681
33	37.6	33.519	4.081
34	37.7	33.312	4.388
34	36.1	33.312	2.788
35	33.6	33.105	0.495
35	34.2	33.105	1.095
36	26.8	32.898	−6.098
36	27.4	32.898	−5.498

c. The residual plot reveals a quadratic upside down (U-shaped) trend. For small and large values of tire pressure, x, the residuals are negative, while for intermediate values of x, the residuals are positive. The quadratic trend suggests the second-order model:

$$E(y) = \beta_0 + \beta_1 x + \beta_2 x^2$$

will provide a better fit than the straight line model.

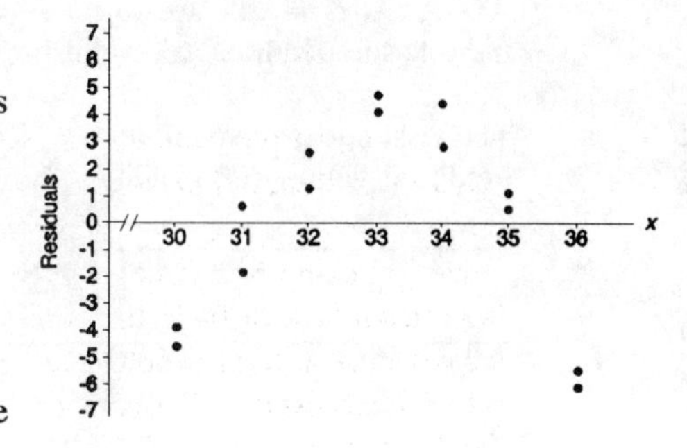

d. The SAS output is:

```
DEP VARIABLE: Y
                              ANALYSIS OF VARIANCE

                               SUM OF                MEAN
SOURCE           DF           SQUARES              SQUARE       F VALUE       PROB>F

MODEL             2         172.40881         86.20440476        70.615       0.0001
ERROR            11       13.42833334          1.22075758
C TOTAL          13         185.83714

           ROOT MSE          1.104879            R-SQUARE        0.9277
           DEP MEAN          33.51429            ADJ R-SQ        0.9146
               C.V.           3.29674

                              PARAMETER ESTIMATES

                          PARAMETER            STANDARD      T FOR HO:
VARIABLE         DF        ESTIMATE               ERROR    PARAMETER=0     PROB>|T|

INTERCEP          1     -1051.10833         92.61766009        -11.349       0.0001
X                 1     66.18571428          5.62799247         11.760       0.0001
XSQ               1     -1.00595238          0.08524326        -11.801       0.0001
```

The least squares regression equation is $\hat{y} = -1051.108 + 66.186x - 1.006x^2$.

To determine if the quadratic term has improved model adequacy, test:

H_0: $\beta_2 = 0$
H_a: $\beta_2 \neq 0$

Test statistic: $t = -11.801$

Conclusion: A p-value of .0001 indicates that we should reject H_0 for any α greater than .0001. At $\alpha = .05$, we reject H_0 and conclude that there is sufficient evidence to indicate the quadratic term did improve model adequacy.

7.5 a. The least-squares equation, $\hat{y} = 9.782 + 1.871x_1 + 1.278x_2$, is used to obtain the predicted values and residuals shown below:

x_1	x_2	Observed Value y	Predicted Value $\hat{y}$	Residual $(y - \hat{y})$
23	5	60.0	59.205	0.795
11	2	32.7	32.919	−0.219
20	9	57.7	58.704	−1.004
17	3	45.5	45.423	0.077
15	8	47.0	48.071	−1.071
21	4	55.3	54.185	1.115
24	7	64.5	63.632	0.868
13	6	42.6	41.773	0.827
19	7	54.5	54.277	0.223
25	2	57.5	59.113	−1.613

b. The plot of the residuals versus x_1 is shown at right.

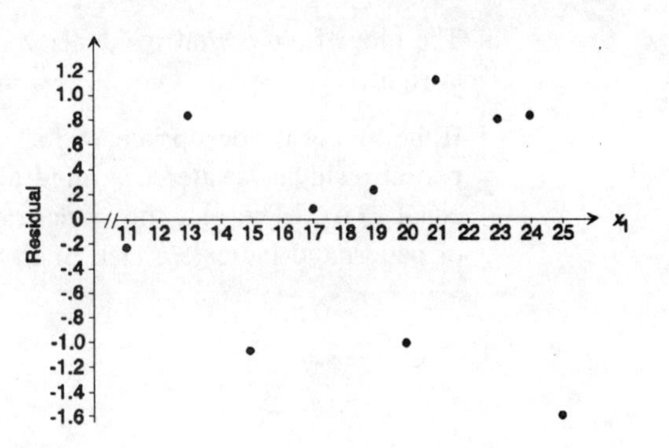

There does not appear to be a trend in the residual plot versus x_1 indicating that the linear x_1-term is probably adequate in the model.

c. The plot of the residuals versus x_2 is shown at right.

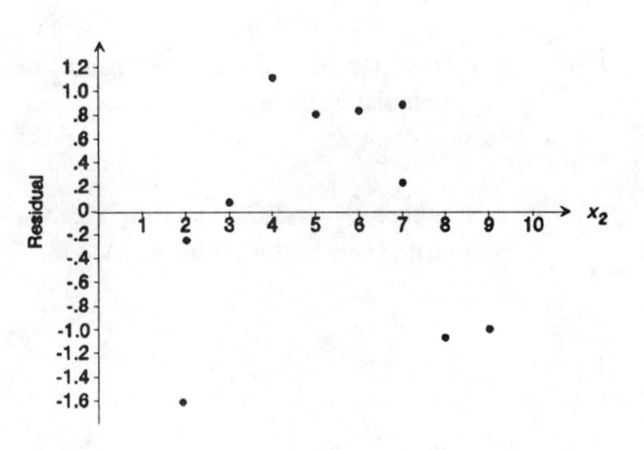

The residual plot versus x_2 reveals a quadratic trend. High and low condition ratings (x_2) tend to have negative residuals while intermediate condition ratings tend to have positive residuals. This trend indicates that perhaps a quadratic term in x_2 would be more appropriate.

In summary, parts (b) and (c) indicate that the model $E(y) = \beta_0 + \beta_1 x_1 + \beta_2 x_2 + \beta_3 x_2^2$ may be more appropriate than the first-order model.

d. The partial residuals, $\hat{\epsilon}^*$ for x_1 will be calculated using:

$$\hat{\epsilon}^* = \hat{\epsilon} + \hat{\beta}_1 x_1$$

where $\hat{\beta}_1 = 1.871$ and the residuals, $\hat{\epsilon}$, are given in the table in (a).

x_1	$\hat{\epsilon}$	$\hat{\epsilon}^*$
23	0.795	43.828
11	−0.219	20.362
20	−1.004	36.416
17	0.077	31.884
15	−1.071	26.994
21	1.115	40.406
24	0.868	45.772
13	0.827	25.150
19	0.223	35.772
25	−1.613	45.162

The plot of the partial residuals versus x_1 is shown at right.

If the model is appropriate, we should see the partial residuals scattered around a line with slope equal to $\hat{\beta}_1$. The plot shows no unusual deviations or patterns to indicate a lack of fit for x_1.

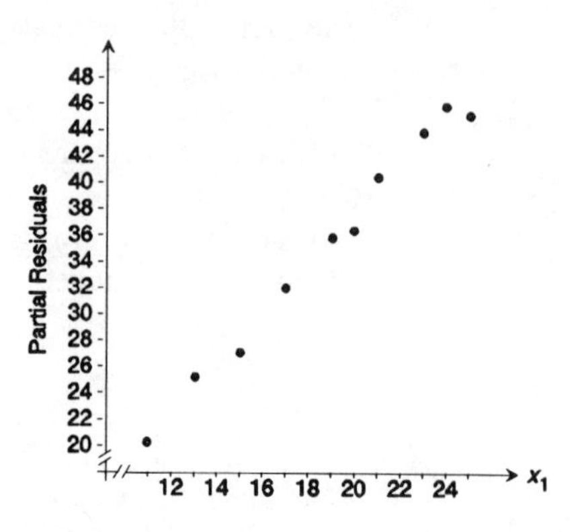

e. The partial residuals, $\hat{\epsilon}^*$, for x_2 will be calculated using:

$$\hat{\epsilon}^* = \hat{\epsilon} + \hat{\beta}_2 x_2$$

where $\hat{\beta}_2 = 1.278$ and the residuals, $\hat{\epsilon}$, are given in the table in (a).

x_2	$\hat{\epsilon}$	$\hat{\epsilon}^*$
5	0.795	7.185
2	−0.219	2.337
9	−1.004	10.498
3	0.077	3.911
8	−1.071	9.153
4	1.115	6.227
7	0.868	9.814
6	0.827	8.495
7	0.223	9.169
2	−1.613	0.943

The plot of the partial residuals versus x_2 is shown at right.

The plot of the partial residuals versus x_2 shows a quadratic trend indicating that a curvature term in x_2 is necessary in the model.

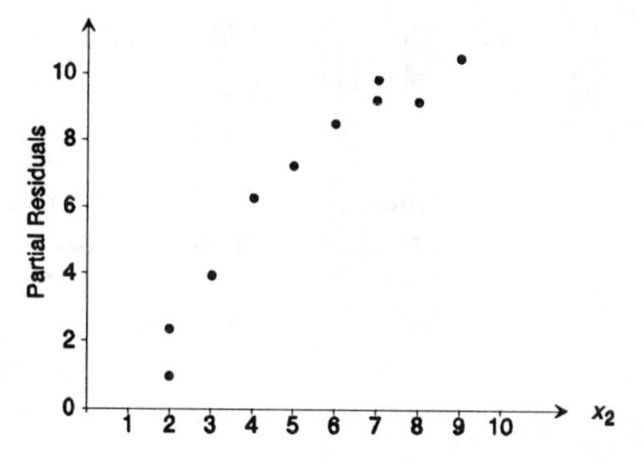

Notice that the results in parts (c) and (d) are similar to the results in parts (a) and (b); that is, that a more appropriate model may be:

$$E(y) = \beta_0 + \beta_1 x_1 + \beta_2 x_2 + \beta_3 x_2^2$$

7.7 a. $-35.1527 + 1.8473 + \cdots + (-3.5334) = 0$

b. The plot of the residuals is:

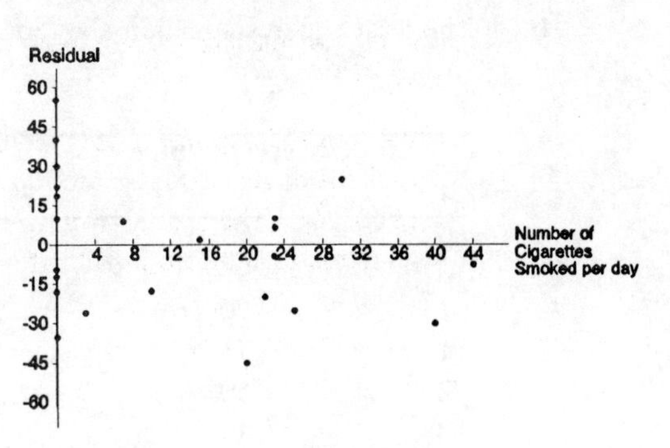

c. No quadratic pattern is discernable in the plot above. There is one value (Residual = 57.3721) that might be an outlier.

We would carefully examine observation #24 to determine if it has been coded correctly.

7.9 $3(s) = 3(4.154) = 12.462$. There are no points that lie above 12.462 or below -12.462. Thus, there are no outliers. The quadratic trend indicates an x^2 term should be added to the model.

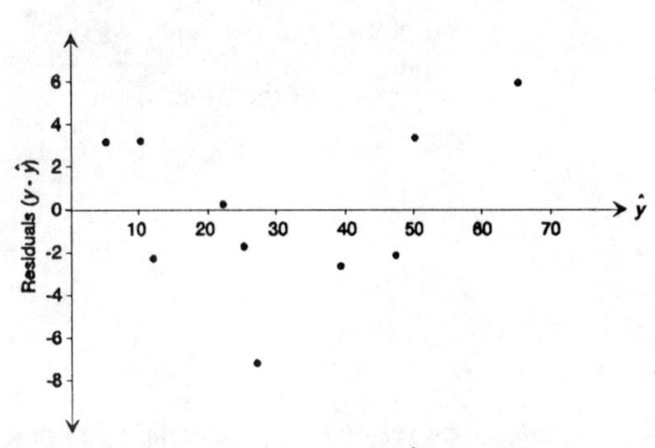

7.11 a. The residual plot indicates that the spread of the residuals increases as $\hat{y}$ increases. This indicates that the assumption of equal variances of the residuals in invalid.

b. The plot of the residuals look very similar to that for a Poisson distribution. The variance stabilizing transformation would be $y^* = \sqrt{y}$

7.13 a. Preliminary calculations are shown below:

$n = 10$ $\quad SS_{xx} = 20$ $\quad \hat{\beta}_1 = -.214$

$\sum x = 20$ $\quad SS_{xy} = -4.28$ $\quad \hat{\beta}_0 = .94$

$\sum y = 5.12$

$\sum x^2 = 60$ $\quad \bar{x} = 2$

$\sum y^2 = 3.5648$ $\quad \bar{y} = .512$

$\sum xy = 5.96$

The least squares prediction equation is $\hat{y} = .94 - .214x$.

b. The least squares equation above is used to obtain the predicted values and residuals shown below:

x	Observed Value y	Predicted Value $\hat{y}$	Residual $(y - \hat{y})$
0	.94	.940	0
0	.96	.940	.020
1	.70	.726	−.026
1	.76	.726	.034
2	.60	.512	.088
2	.40	.512	−.112
3	.24	.298	−.058
3	.30	.298	.002
4	.12	.084	.036
4	.10	.084	.016

The plot of the residuals versus $\hat{y}$ is shown at right:

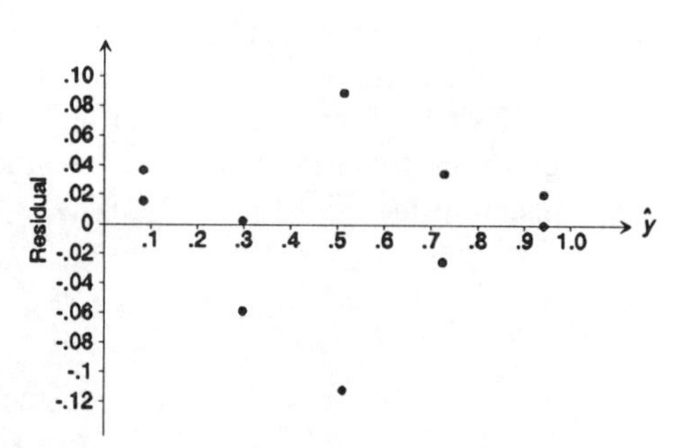

c. The residual plot reveals a football-shaped pattern (smaller variances for small and large values of $\hat{y}$ and larger variances for intermediate values of $\hat{y}$). This indicates heteroscedasaticity. When the response variable, y, is a proportion or percentage, the data often fail to be homoscedastic.

d. An appropriate variance-stabilizing transformation when the responses are generated by a binomial experiment is:

$$y^* = \sin^{-1}\sqrt{y}$$

Then fit the model. $y^* = \beta_0 + \beta_1 x + \epsilon$

e. Preliminary calculations are shown:

$\sum y^* = 8.0806$ $\quad SS_{xy^*} = -4.9928$

$\sum xy^* = 11.1684$ $\quad \hat{\beta}_1 = -.2496$

$\bar{y}^* = .80806$ $\quad \hat{\beta}_0 = 1.307$

The least squares regression equation is $\hat{y}^* = 1.307 - .2496x$.

x	$y^* =$ (in radians)
0	1.3233
0	1.3694
1	.9912
1	1.0588
2	.8861
2	.6847
3	.5120
3	.5796
4	.3537
4	.3218

The least squares equation above is used to obtain the predicted values and residuals shown below:

x	Observed Value y^*	Predicted Value $\hat{y}^*$	Residual $(y^* - \hat{y}^*)$
0	1.3233	1.3070	.0163
0	1.3694	1.3070	.0624
1	.9912	1.0574	−.0662
1	1.0588	1.0574	.0014
2	.8861	.8078	.0783
2	.6847	.8078	−.1231
3	.5120	.5582	−.0462
3	.5796	.5582	−.0214
4	.3537	.3086	.0451
4	.3218	.3086	.0132

The plot of the residuals versus $\hat{y}^*$ is shown at right:

This plot still reveals the football-shaped pattern. The variances may have been equalized somewhat, but this transformation does not seem to have given us the desired effect.

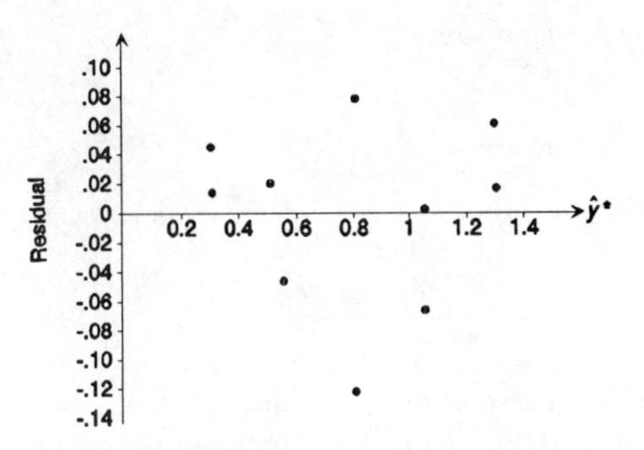

7.15 The stem-and-leaf display indicates that the data are fairly mound-shaped. There is no evidence that the data are not normal. The box plot shows that the data are fairly symmetrical and mound-shaped. Again, there is no evidence that the data are not normal. The normal probability plot is a fairly straight line. Again, this indicates that there is no evidence that the data are not normal. Based on all three of these graphs, our assumption of normality seems to be valid.

7.17 The residuals for 7.14 are as follows:

Property	Residual	Property	Residual
1	−29.29	11	38.26
2	−8.05	12	18.08
3	−51.55	13	21.91
4	.16	14	.46
5	23.69	15	31.14
6	−1.09	16	−13.92
7	−29.68	17	6.07
8	−11.83	18	−9.14
9	41.37	19	−9.08
10	−3.57	20	−13.92

We can construct a relative frequency histogram using the following classes:

Class	Frequency	Relative Frequency
−52 to −37	1	.05
−37 to −22	2	.10
−22 to −7	6	.30
−7 to 8	5	.25
8 to 23	2	.10
23 to 38	2	.10
38 to 53	2	.10
	20	1.00

The residuals do not appear to deviate drastically from normality. It is probably safe to say the assumption of normality is not violated.

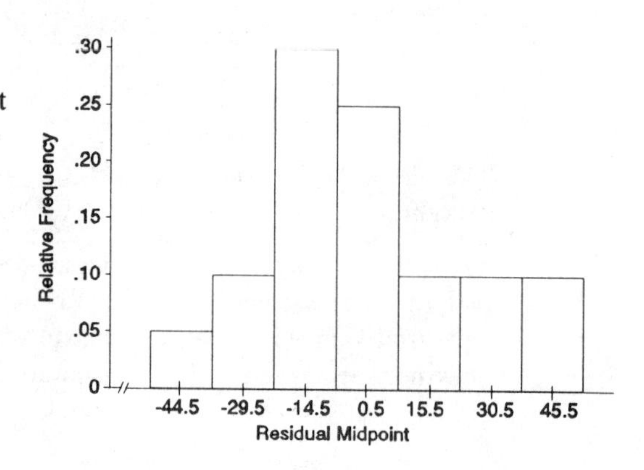

7.19 The plot of the residuals versus $\hat{y}$ is shown at right. To look for outliers, we compute:

$$3s = 3\sqrt{\text{MSE}} = 3\sqrt{.1267} = 1.07$$

Looking down the fourth column (residual) of the table shown in the solution to Exercise 7.1, we see that none of the residuals exceed $3s$, or 1.07, in absolute value. No outliers are present in the data.

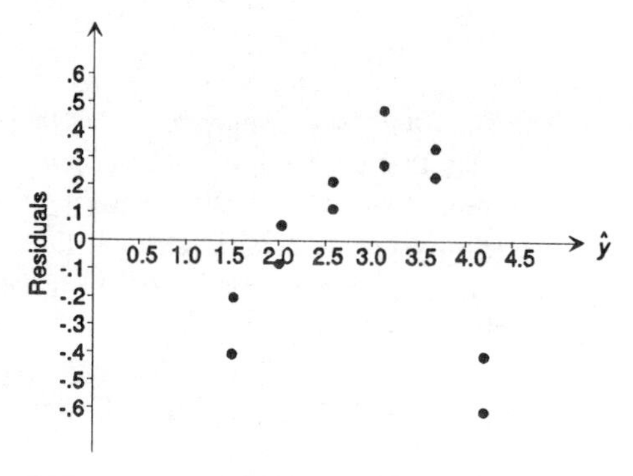

7.21 a. To determine if the model is adequate, we test:

H_0: $\beta_1 = 0$
H_a: $\beta_1 \neq 0$

Test statistic: $F = 21.772$

The p-value associated with this F is $p = .0001$. Since this p-value is so small, we reject H_0. There is sufficient evidence to indicate the model is adequate.

b. The output from fitting the model using SAS is:

Model: MODEL1
Dependent Variable: PCB85

Analysis of Variance

Source	DF	Sum of Squares	Mean Square	F Value	Prob>F
Model	1	466788.58722	466788.58722	28.670	0.0001
Error	34	553577.79946	16281.69998		
C Total	35	1020366.3867			

Root MSE	127.59976	R-square	0.4575
Dep Mean	95.34083	Adj R-sq	0.4415
C.V.	133.83538		

Parameter Estimates

Variable	DF	Parameter Estimate	Standard Error	T for H0: Parameter=0	Prob > |T|
INTERCEP	1	72.973773	21.67301366	3.367	0.0019
PCB84	1	0.040649	0.00759164	5.354	0.0001

Obs	BAY	PCB84	Dep Var PCB85	Predict Value	Residual
1	Casco	95.28	77.5500	76.8468	0.7032
2	Merrmack	52.97	29.2300	75.1269	-45.8969
3	Salem	533.58	403.1	94.6631	308.4
4	Boston	17104.86	736.0	768.3	-32.2627
5	BUZZARDS	308.46	192.2	85.5122	106.6
6	Narragan	159.96	220.6	79.4759	141.1
7	ELongIsl	10	8.6200	73.3803	-64.7603
8	WLONGISL	234.43	174.3	82.5030	91.8070
9	RARITAN	443.89	529.3	91.0173	438.3
10	DELAWARE	2.5	130.7	73.0754	57.5946
11	LChesapk	51	39.7400	75.0469	-35.3069
12	Pamilico	0	0	72.9738	-72.9738
13	Charlest	9.1	8.4300	73.3437	-64.9137
14	Sapelo	0	0	72.9738	-72.9738
15	STJOHNS	140	120.0	78.6646	41.3754
16	Tampa	0	0	72.9738	-72.9738
17	Apalach	12	11.9300	73.4616	-61.5316
18	Mobile	0	0	72.9738	-72.9738
19	RoundIsl	0	0	72.9738	-72.9738
20	MissRiv	34	30.1400	74.3558	-44.2158
21	Baratara	0	0	72.9738	-72.9738
22	SanAnton	0	0	72.9738	-72.9738
23	CorpusCh	0	0	72.9738	-72.9738
24	SDiegoBa	6.74	9.3000	73.2477	-63.9477
25	DabaPt	7.06	5.7400	73.2608	-67.5208
26	SealBch	46.71	46.4700	74.8725	-28.4025
27	SanPedro	159.56	176.9	79.4597	97.4403
28	SantaMon	14	13.6900	73.5429	-59.8529
29	Bodega	4.18	4.8900	73.1437	-68.2537
30	Coos	3.19	6.6000	73.1034	-66.5034
31	Columbia	8.77	6.7300	73.3303	-66.6003
32	Nisquall	4.23	4.2800	73.1457	-68.8657
33	Commence	20.6	20.5000	73.8111	-53.3111
34	Elliot	329.97	414.5	86.3866	328.1
35	Lutak	5.5	5.8000	73.1973	-67.3973
36	Nahku	6.6	5.0800	73.2421	-68.1621

By eliminating San Diego Harbor, the model adequacy is a little better. The R^2 value has increased from .3835 to .4575. The value of s has decreased from 145.7252 to 127.59976. The largest residual is now 438. However, we should note that in the original model, there were several fairly large residuals besides just the one from San Diego Harbor. Salem (296.5), Raritan (426.3), and Elliot (316.1) all had fairly large residuals. By eliminating only one, the fit of the model will not be that much better.

c. H_0: $\beta_1 = 0$
H_a: $\beta_1 \neq 0$

Test statistic: $F = 251.172$

The p-value associated with this F is $p = .0001$. Since this p-value is so small, we reject H_0. There is sufficient evidence to indicate the model is adequate.

The R^2 value for this model is .8777. This is much greater than either of the other two models. The residual for San Diego Harbor is now .7073. The standard deviation for this model is .76132. Thus, the residual for San Diego Harbor is less than one standard deviation from the mean. It is no longer an outlier.

7.23 To look for outliers, we compute $3s = 3\sqrt{\text{MSE}} = 3\sqrt{.036} = .57$

None of the residuals on the SAS residual plot exceed $3s$, or .57, in absolute value. No outliers are present in the data.

7.25 a. From the printout, the estimate of the standard deviation is Root MSE = 24.68155. The largest residual in magnitude is 57.3721, corresponding to observation 24. This residual is only 2.32 standard deviations from the mean of 0. No residual is more than 3 standard deviations from the mean. Thus, it appears that there are no outliers.

b. Leverage values for each observation are given under the column heading HAT DIAG H. The largest leverage value is .2590, which is associated with the 24th observation. The average leverage for all $n = 25$ observations is:

$$\bar{h} = \frac{k + 1}{n} = \frac{2}{25} = .08$$

.2590 is much greater than two times the average leverage value, 2(.08) = .16. This implies that observation 24 is an influential observation. Also, observation 5 has a leverage value of .2065. This value is also much larger than .16. Thus, it appears that observation 5 is an influential observation.

Another measure of influence is Cook's D. Cook's distance is a function of both the leverage and the magnitude of the residual. For this example, the largest Cook's D is associated with observation 24 and is 1.275. This value is quite large compared with all the other values (the next largest value is .221 associated with observation 5). D_{24} falls in the 70th percentile of the F distribution with $\nu_1 = k + 1 = 2$ and $\nu_2 = n - (k + 1) = 25 - 2 = 23$ degrees of freedom. This implies observation 24 has substantial influence on the estimates of the model parameters.

The Studentized deleted residual, under the column heading RSTUDENT, is also used to detect influential observations. The Studentized deleted residual has an approximate t distribution with $(n - 1) - (k + 1) = 22$ df. The largest value under the RSTUDENT column is 3.1959, again associated with observation 24. The probability of getting a t value greater than 3.1959 is between .001 and .005. This would be extremely rare. Again, this implies that observation 24 is highly influential. All other RSTUDENT values lie between -1.6 and 1.6.

The DIFFTS column gives the difference between the predicted value when all 24 observations are used and when the ith observation is omitted, divided by its standard error. For this example, the DIFFTS value with the largest magnitude is 1.8896, associated with observation 24. The next largest value in magnitude is $-.6762$, associated with observation 5. The DIFFTS value of 1.8896 is almost three times as large as the next largest. This implies that observation 24 is highly influential.

The final measure used to measure influence is the DFBETAS. These measures give the difference in the parameter estimates with all observations and the estimates without the ith observation, divided by the standard error. The largest DFBETA - Intercept in magnitude is associated with observation 24 and is $-.6678$. The next largest value in magnitude is .5356, associated with observation 22. Because the largest value is not that different from the next largest value, no inference is made. The largest DFBETA - X in magnitude is associated with observation 24 and is 1.7376. The next largest value in magnitude is $-.6072$, associated with observation 5. Because the largest value is much larger than the next largest value, it implies that observation 24 is highly influential.

In summary, observation 24 is inferred to be highly influential using several measures. Observation 5 is inferred to be influential using only the leverage measure. Thus, we conclude that only observation 24 is highly influential.

7.27 a. For Bank 1, $R^2 = .914$. Thus, 91.4% of the variation in the deposit share values for Bank 1 is explained by the model containing expenditures on promotion related activities, expenditures on service related activities, and expenditures on distribution related activities.

For Bank 2, $R^2 = .721$. Thus, 72.1% of the variation in the deposit share values for Bank 2 is explained by the model containing expenditures on promotion related activities, expenditures on service related activities, and expenditures on distribution related activities.

For Bank 3, $R^2 = .926$. Thus, 92.6% of the variation in the deposit share values for Bank 3 is explained by the model containing expenditures on promotion related activities, expenditures on service related activities, and expenditures on distribution related activities.

For Bank 4, $R^2 = .827$. Thus, 82.7% of the variation in the deposit share values for Bank 4 is explained by the model containing expenditures on promotion related activities, expenditures on service related activities, and expenditures on distribution related activities.

For Bank 5, $R^2 = .270$. Thus, 27.0% of the variation in the deposit share values for Bank 5 is explained by the model containing expenditures on promotion related activities, expenditures on service related activities, and expenditures on distribution related activities.

For Bank 6, $R^2 = .616$. Thus, 61.6% of the variation in the deposit share values for Bank 6 is explained by the model containing expenditures on promotion related activities, expenditures on service related activities, and expenditures on distribution related activities.

For Bank 7, $R^2 = .962$. Thus, 96.2% of the variation in the deposit share values for Bank 7 is explained by the model containing expenditures on promotion related activities, expenditures on service related activities, and expenditures on distribution related activities.

For Bank 8, $R^2 = .495$. Thus, 49.5% of the variation in the deposit share values for Bank 8 is explained by the model containing expenditures on promotion related activities, expenditures on service related activities, and expenditures on distribution related activities.

For Bank 9, $R^2 = .500$. Thus, 50.0% of the variation in the deposit share values for Bank 9 is explained by the model containing expenditures on promotion related activities, expenditures on service related activities, and expenditures on distribution related activities.

b. To test for the overall adequacy of the models, we test:

H_0: $\beta_1 = \beta_2 = \beta_3 = 0$
H_a: At least one $\beta_1 \neq 0$, $i = 1, 2, 3$

For Bank 1, the p-value for the global F test statistic is $p = .000$. Since this p-value is so small, we reject H_0. There is sufficient evidence to indicate that the overall model is adequate at $\alpha = .05$.

For Bank 2, the p-value for the global F test statistic is $p = .004$. Since this p-value is so small, we reject H_0. There is sufficient evidence to indicate that the overall model is adequate at $\alpha = .05$.

For Bank 3, the p-value for the global F test statistic is $p = .000$. Since this p-value is so small, we reject H_0. There is sufficient evidence to indicate that the overall model is adequate at $\alpha = .05$.

For Bank 4, the p-value for the global F test statistic is $p = .000$. Since this p-value is so small, we reject H_0. There is sufficient evidence to indicate that the overall model is adequate at $\alpha = .05$.

For Bank 5, the p-value for the global F test statistic is $p = .155$. Since this p-value is not so small, we do not reject H_0. There is insufficient evidence to indicate that the overall model is adequate at $\alpha = .05$.

For Bank 6, the p-value for the global F test statistic is $p = .012$. Since this p-value is so small, we reject H_0. There is sufficient evidence to indicate that the overall model is adequate at $\alpha = .05$.

For Bank 7, the p-value for the global F test statistic is $p = .000$. Since this p-value is so small, we reject H_0. There is sufficient evidence to indicate that the overall model is adequate at $\alpha = .05$.

For Bank 8, the p-value for the global F test statistic is $p = .014$. Since this p-value is so small, we reject H_0. There is sufficient evidence to indicate that the overall model is adequate at $\alpha = .05$.

For Bank 9, the p-value for the global F test statistic is $p = .011$. Since this p-value is so small, we reject H_0. There is sufficient evidence to indicate that the overall model is adequate at $\alpha = .05$.

c. To determine if positive autocorrelation is present, we test:

H_0: No residual correlation
H_a: Residual correlation exists

For $\alpha = .10$, the rejection region is $d < d_{L,\alpha/2}$ or $(4 - d) < d_{L,\alpha/2}$. From Table 9, Appendix C, for $n = 20$ and $k = 3$, $d_{L,.05} = 1.00$. The rejection region is $d < 1.00$ or $(4 - d) < 1.00$. If $d > d_{U,.05}$ or $(4 - d) > d_{U,.05}$, then H_0 is not rejected. If $d_{L,.05} < d < d_{U,.05}$ or $d_{L,.05} < (4 - d) < d_{U,.05}$, no conclusion is reached. From Table 9, Appendix C, with $n = 20$ and $k = 3$, $d_{U,.05} = 1.68$.

For Bank 1, $d = 1.3$ and $4 - d = 4 - 1.3 = 2.7$. Since $1.00 < d = 1.3 < 1.68$, no conclusion can be reached.

For Bank 2, $d = 3.4$ and $4 - d = 4 - 3.4 = 0.6$. Since $4 - d = 0.6 < 1.00$, H_0 is rejected. There is sufficient evidence to indicate autocorrelation is present.

For Bank 3, $d = 2.7$ and $4 - d = 4 - 2.7 = 1.3$. Since $1.00 < 4 - d = 1.3 < 1.68$, no conclusion can be reached.

For Bank 4, $d = 1.9$ and $4 - d = 4 - 1.9 = 2.1$. Since $d = 1.9 > 1.68$, H_0 is not rejected. There is insufficient evidence to indicate autocorrelation is present.

For Bank 5, $d = 0.85$ and $4 - d = 4 - 0.85 = 3.15$. Since $d = 0.85 < 1.00$, H_0 is rejected. There is sufficient evidence to indicate autocorrelation is present.

For Bank 6, $d = 1.8$ and $4 - d = 4 - 1.8 = 2.2$. Since $d = 1.8 > 1.68$, H_0 is not rejected. There is insufficient evidence to indicate autocorrelation is present.

For Bank 7, $d = 2.5$ and $4 - d = 4 - 2.5 = 1.5$. Since $1.00 < 4 - d = 1.5 < 1.68$, no conclusion can be reached.

For Bank 8, $d = 2.3$ and $4 - d = 4 - 2.3 = 1.7$. Since $4 - d = 1.7 > 1.68$, H_0 is not rejected. There is insufficient evidence to indicate autocorrelation is present.

For Bank 9, $d = 1.1$ and $4 - d = 4 - 1.1 = 2.9$. Since $1.00 < d = 1.1 < 1.68$, no conclusion can be reached.

7.29 a. The plot of the residuals against time is as follows:

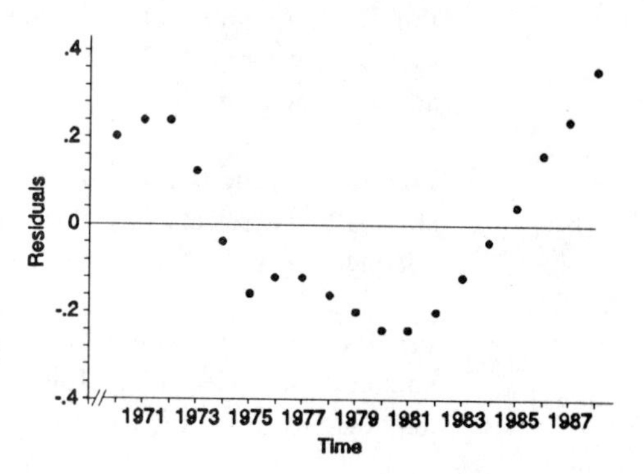

The first 4 residuals are positive, the next 11 residuals are negative, while the last 4 residuals are positive. There is a tendency for long positive and long negative runs. This implies that there is positive residual correlation. One can account for this by the fact that the purchasing power of a dollar in one year is not independent of the purchasing power of a dollar in the year preceding or following that year.

b. The Durbin-Watson d statistic is $d = .173$.

H_0: No residual correlation
H_a: Positive residual correlation

Test statistic: $d = .173$

Rejection region: $\alpha = .05$, $k = 1$, $n = 19$, $d_{L,.05} = 1.18$, $d_{U,.05} = 1.40$
Reject H_0 if $d < 1.18$,
Do not reject H_0 if $d > 1.40$
Inconclusive if $1.18 \le d \le 1.40$

Conclusion: Since $d = .173 < 1.18$, reject H_0 at $\alpha = .05$. There is sufficient evidence to indicate that the residuals are positively correlated.

7.31 a. H_0: No residual correlation
H_a: Positive residual correlation

Test statistic: $d = \dfrac{\sum(\hat{\varepsilon}_t - \hat{\varepsilon}_{t-1})^2}{\sum \hat{\varepsilon}_t^2} = 2.09$

Rejection region: $\alpha = .05$, $n = 40$, $k = 2$, $d_{L,.05}$, $= 1.39$, $d_{U,.05} = 1.60$
Reject H_0 if $d < 1.39$,
Do not reject H_0 if $d > 1.60$
Test is inconclusive if $1.39 \le d \le 1.60$

Conclusion: Do not reject H_0 at $\alpha = .05$. There is insufficient evidence to indicate that the residuals are positively correlated. [Remember, however, that the Durbin-Watson test is designed to detect first-order autocorrelation. The residuals may have a more complex (higher than first-order) autocorrelation pattern.]

b. H_0: No residual correlation
H_a: Positive residual correlation

Test statistic: $d = \dfrac{\sum(\hat{\varepsilon}_t - \hat{\varepsilon}_{t-1})^2}{\sum \hat{\varepsilon}_t^2} = 1.07$

Rejection region: $\alpha = .05$, $n = 40$, $k = 2$, $d_{L,.05} = 1.39$, $d_{U,.05} = 1.60$
Reject H_0 if $d < 1.39$,
Do not reject H_0 if $d > 1.60$
Test is inconclusive if $1.39 \le d \le 1.60$

Conclusion: Reject H_0 at $\alpha = .05$. There is sufficient evidence to indicate that the residuals are positively correlated.

7.33 a. The curvilinear trend implies a misspecified model. A quadratic term needs to be added.

b. The spread of the residuals increasing as $\hat{y}$ increases implies unequal variances.

c. A residual lying more than 3 standard deviations from the mean implies an outlier is present.

d. The spread of the residuals increasing and then decreasing as $\hat{y}$ increases implies unequal variances.

e. The histogram indicates nonnormal errors because it is skewed to the right.

7.35 a. H_0: $\beta_1 = \beta_2 = \cdots = \beta_7 = 0$
H_a: At least one coefficient is nonzero

Test statistic: $F = \dfrac{R^2/k}{(1 - R^2)/[n - (k + 1)]} = 217.23$

Rejection region: $\alpha = .05$, $\nu_1 = 7$, $\nu_2 = 20$, $F_{.05} = 2.51$
Reject H_0 if $F > 2.51$

Conclusion: Reject H_0. There is sufficient evidence (at $\alpha = .05$) to indicate that the model is useful for predicting annual traffic fatalities.

b. H_0: No residual correlation
H_a: Positive residual correlation

Test statistic: $d = \dfrac{\sum(\hat{\varepsilon}_t - \hat{\varepsilon}_{t-1})^2}{\sum \hat{\varepsilon}_t^2} = 1.97$

Rejection region: $\alpha = .05$, $n = 28$, $k = 7$, $d_{L,.05} \approx 1.03$, $d_{U,.05} \approx 1.85$
Reject H_0 if $d < 1.03$
Do not reject H_0 if $d > 1.85$
Test is inconclusive if $1.03 \le d \le 1.85$

Conclusion: Do not reject H_0 at $\alpha = .05$. There is insufficient evidence to indicate that the regression results are positively correlated.

7.37 a. The plot of the residuals versus x_1 is shown at right:

The residual plot reveals a football-shaped pattern. This indicates heteroscedasticity (nonconstant variance).

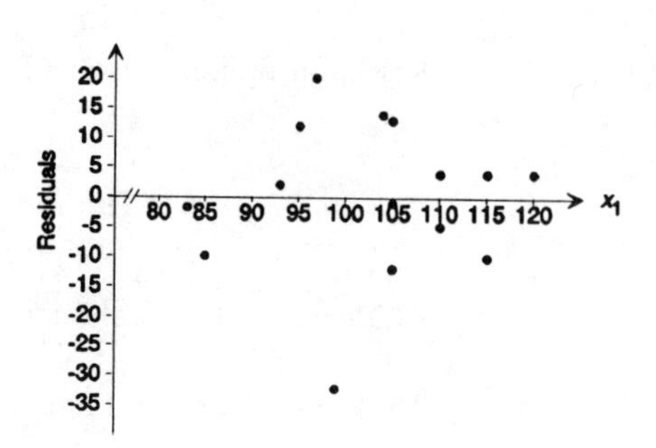

The plot of the residuals versus x_2 is shown at right:

The residual plot reveals a fan-shaped pattern with larger variances for smaller values of x_2 and smaller variances for larger values of x_2. This also indicates heteroscedasticity.

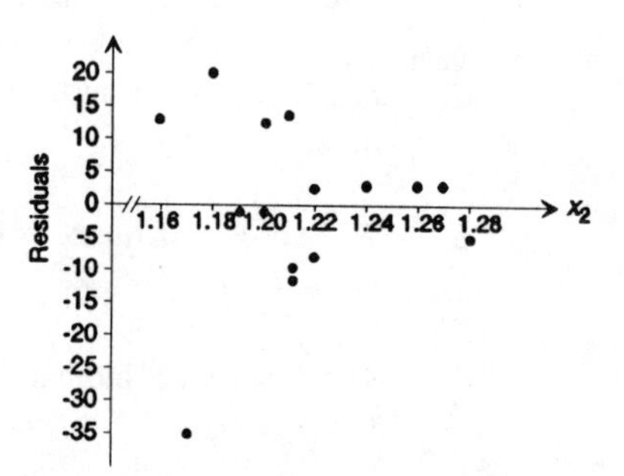

b. The partial residuals, $\hat{\epsilon}^*$, for x_1 will be calculated using:

$$\hat{\epsilon}^* = \hat{\epsilon} + \hat{\beta}_1 x_1$$

where $\hat{\beta}_1 = -0.7898$ and the residuals, $\hat{\epsilon}$, are given in the problem.

x_1	$\hat{\epsilon}$	$\hat{\epsilon}^*$
93	1.27	−72.18
95	−12.63	−62.40
85	−10.27	−77.40
83	−0.13	−65.68
95	20.72	−54.31
103	13.76	−67.59
100	−34.47	−113.45
105	−0.05	−82.98
105	11.95	−70.98
105	−11.91	−94.84
115	−9.35	−100.18
120	3.88	−90.90
115	4.21	−86.62
110	4.32	−82.56
110	−6.54	−93.42

The plot of the partial residuals versus x_1 is shown at right:

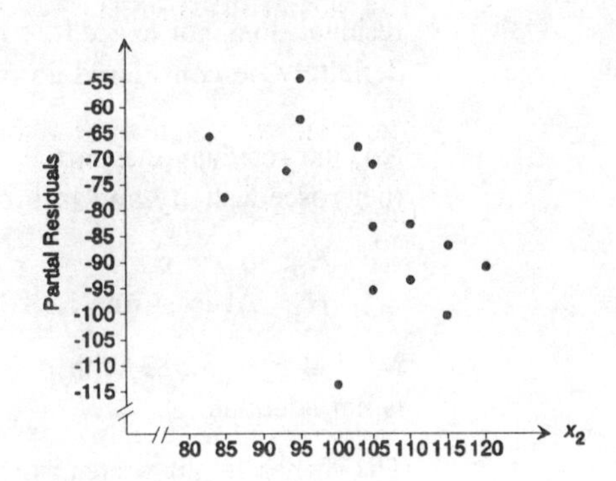

If the model is appropriate, we should see the partial residuals scattered around a line with slope equal to $\hat{\beta}_1$. The plot does show a large amount of deviation from this line indicating a lack of fit for x_1.

c. The partial residuals, $\hat{\epsilon}^*$, for x_2 will be calculated using:

$$\hat{\epsilon}^* = \hat{\epsilon} + \hat{\beta}_2 x_2$$

where $\hat{\beta}_2 = 285.969$ and the residuals are given in the problem.

x_1	$\hat{\epsilon}$	$\hat{\epsilon}^*$
1.22	1.27	350.15
1.21	12.63	358.65
1.21	−10.27	335.75
1.19	−0.13	340.17
1.18	20.72	358.16
1.16	13.76	345.48
1.17	−34.47	300.11
1.20	−0.05	343.11
1.20	11.95	355.11
1.21	−11.91	334.11
1.22	−9.35	339.53
1.24	3.88	358.48
1.26	4.21	364.53
1.27	4.32	367.50
1.28	−6.54	359.50

The plot of the partial residuals versus x_2 is shown at right:

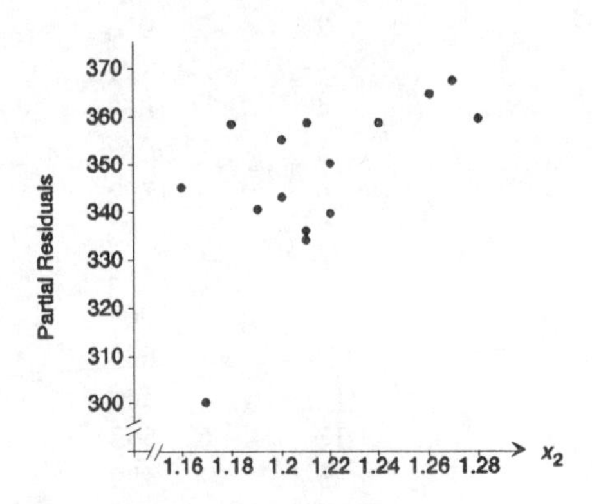

If the model is appropriate, we should see the partial residuals scattered around a line with slope equal to $\hat{\beta}_2$. The plot shows large deviations from this line indicating a lack of fit for x_2.

d. To look for outliers, we compute $2s = 2\sqrt{\text{MSE}} = 2\sqrt{282.19383} = 33.6$

Looking at the plots in part (a), only one observation (Nov 87) has a residual exceeding $2s$ or 33.6 in absolute value. This could be considered an outlier, but notice that the residual does not exceed 3s in absolute value in which case the observation would definitely be considered an outlier.

e. No, the residual and partial residual plots all indicate problems with the model (heteroscedasticity and lack of fit). Also, for

H_0: $\beta_1 = \beta_2 = \cdots = \beta_5 = 0$
H_a: At least one coefficient is nonzero

we find $F = 3.383$ with $p = .0538$ (from printout in text). This also indicates the model is not adequate.

The owner might want to try a model with interaction terms. Also, because the plots indicate non-constant variance, a variance stabilizing transformation is indicated.

7.39 a. Preliminary calculations are shown below:

$n = 13$ | $SS_{tt} = 182$ | $\hat{\beta}_1 = -64.1044$
$\sum t = 91$ | $SS_{ty_t} = -11667$ | $\hat{\beta}_0 = 1604.81$
$\sum y_t = 15029$ | $\bar{t} = 7$
$\sum t^2 = 819$ | $\bar{y}_t = 1156.0769$
$\sum y_t^2 = 21{,}244{,}983$
$\sum ty_t = 93536$

The least squares prediction equation is $\hat{y}_t = 1064.81 - 64.1044t$.

b. The least squares equation above is used to obtain the predicted values and residuals shown in the table:

t	Observed Value y_t	Predicted Value $\hat{y}$	Residual $(y - \hat{y})$
1	767	1504.7	−773.7
2	926	1476.6	−550.6
3	1171	1412.5	−241.5
4	1663	1348.4	314.6
5	2058	1284.3	773.7
6	1892	1220.2	671.8
7	1866	1156.1	709.9
8	1414	1092.0	322.0
9	1067	1027.9	39.1
10	633	963.8	−330.8
11	540	899.7	−359.7
12	553	835.6	−282.6
13	479	771.5	−292.5

The plot of the residuals versus t is shown at right:

There is a tendency for the residuals to have long positive and negative runs. These cycles indicate possible residual correlation.

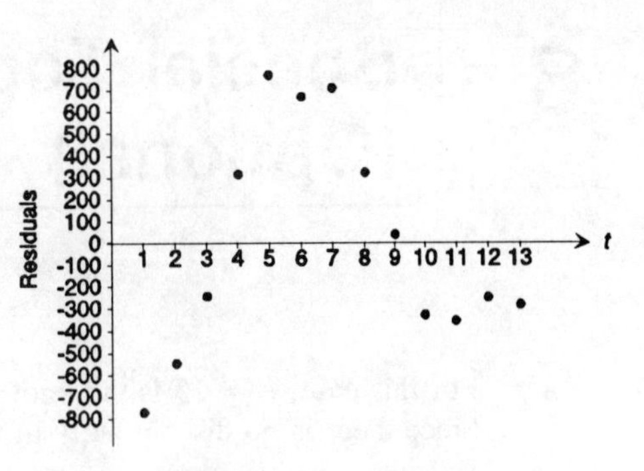

c. Using the regression residuals obtained in part (b), we compute:

$$d = \frac{\sum (\hat{\varepsilon}_t - \hat{\varepsilon}_{t-1})^2}{\sum \hat{\varepsilon}_t^2} = \frac{[-550.6 - (-773.7)]^2 + \cdots + [-292.5 - (-282.6)]^2}{(-773.7)^2 + (-550.6)^2 + \cdots + (-292.5)^2}$$

$$= \frac{1051374.12}{3122397} = .337$$

H_0: No residual correlation
H_a: Positive residual correlation

Test statistic: $d = .337$

Rejection region: $\alpha = .01$, $n = 13$, $k = 1$, $d_{L,.01} \approx 0.81$, $d_{U,.01} \approx 1.07$
Reject H_0 if $d < 0.81$
Do not reject H_0 if $d > 1.07$
Test is inconclusive if $0.81 \le d \le 1.07$

Conclusion: Reject H_0 at $\alpha = .01$. There is sufficient evidence to indicate positive residual correlation.

7.41 a. H_0: No residual correlation
H_a: Positive residual correlation

Test statistic: $d = \dfrac{\sum_{t=2}^{n} (\hat{\varepsilon}_t - \hat{\varepsilon}_{t-1})^2}{\sum_{t=1}^{n} \hat{\varepsilon}_t^2}$

Rejection region: $\alpha = .05$, $k = 2$, $n = 40$, $d_{L,.05} = 1.39$, $d_{U,.05} = 1.60$
Reject H_0 if $d < 1.39$,
Do not reject H_0 if $d > 1.60$
Inconclusive if $1.39 \le d \le 1.60$

b. Test statistic: $d = 1.14$

Conclusion: Since $d = 1.14 < 1.39$, reject H_0 at $\alpha = .05$. There is sufficient evidence to indicate that the residuals are positively correlated.

8 Special Topics in Regression (Optional)

8.1 a. For this case, $k = 15$ is the knot value (i.e., the value of x at which the slope changes). Since there is no discontinuity in the linear relationship, the appropriate model is:

$$E(y) = \beta_0 + \beta_1 x_1 + \beta_2(x_1 - 15)x_2$$

where $x_1 = x$ and $x_2 = \begin{cases} 1 & \text{if } x > 15 \\ 0 & \text{if not} \end{cases}$

b. For $x \leq 15$, $x_2 = 0$, and the model is:

$$E(y) = \beta_0 + \beta_1 x$$

Thus, β_0 = y-intercept and β_1 = slope.

For $x > 15$, $x_2 = 1$, and the model is:

$$E(y) = \beta_0 + \beta_1 x + \beta_2(x - 15) = (\beta_0 - 15\beta_2) + (\beta_1 + \beta_2)x$$

Thus, $(\beta_0 - 15\beta_2)$ = y-intercept and $(\beta_1 + \beta_2)$ = slope.

c. If the two slopes are identical, then we must have $\beta_2 = 0$. Therefore, to test for a difference between the two slopes, we test:

$$H_0\text{: } \beta_2 = 0 \text{ against } H_a\text{: } \beta_2 \neq 0$$

using a t test.

8.3 a. For this case, $k = 320$ is the knot value. Since there is discontinuity in the linear relationship, the appropriate linear model is:

$$E(y) = \beta_0 + \beta_1 x_1 + \beta_2(x_1 - 320)x_2 + \beta_3 x_2$$

where $x_1 = x$ and $x_2 = \begin{cases} 1 & \text{if } x > 320 \\ 0 & \text{if not} \end{cases}$

b. For $x \leq 320$, $x_2 = 0$, and the model is:

$$E(y) = \beta_0 + \beta_1 x$$

Thus, β_0 = y-intercept and β_1 = slope.

For $x > 320$, $x_2 = 1$, and the model is:

$$E(y) = \beta_0 + \beta_1 x + \beta_2(x - 320) + \beta_3 = (\beta_0 - 320\beta_2 + \beta_3) + (\beta_1 + \beta_2)x$$

Thus, $(\beta_0 - 320\beta_2 + \beta_3)$ = intercept and $(\beta_1 + \beta_2)$ = slope.

c. To test for a difference between the two lines, we test H_0: $\beta_2 = \beta_3 = 0$ using a t test.

8.5 a. With no discontinuity and a knot value of $k = 1000$, (i.e., change in slope at $x = 1000$), the appropriate piecewise linear model is:

$$E(y) = \beta_0 + \beta_1 x_1 + \beta_2(x_1 - 1000)x_2$$

where $x_1 = x =$ lot size, and $x_2 = \begin{cases} 1 & \text{if } x > 1000 \\ 0 & \text{if not} \end{cases}$

b. The least squares prediction equation, obtained from the SAS regression procedure, is given by:

$$\hat{y} = 4.02402680 - .002089716x_1 - .001393686(x_1 - 1000)x_2$$

c. H_0: $\beta_1 = \beta_2 = 0$
H_a: At least one coefficient is not 0

Test statistic: $F = 363.44$ (from printout)

Rejection region: From the printout, the p-value associated with $F = 363.44$ is .0001.

Conclusion: Since the p-value is less than $\alpha = .10$, reject H_0. There is sufficient evidence to indicate that the model is adequate for predicting shipping cost, y.

d. The mean increase in shipping cost for every unit increase in lot size is represented by the slope of the $y - x$ line. For lots with 1,000 or fewer units (i.e., for $x \leq 1000$), the slope is represented by β_1 (since $x_2 = 0$). A 90% confidence interval for β_1 is:

$$\hat{\beta}_1 \pm t_{.05} s_{\hat{\beta}_1}$$

where $t_{.05} = 1.782$ is based on 12 df and $s_{\hat{\beta}_1} = .000205206$ is obtained from the SAS printout. Substitution yields:

$$-.002089716 \pm (1.782)(.000205206) \Rightarrow -.002089716 \pm .0003657$$
$$\text{or } (-.0024554, -.001724)$$

Since the interval includes only negative numbers, there is evidence that mean shipping cost for lot sizes of 1,000 or fewer units decreases as the lot size increases.

8.7 a. Some preliminary calculations are:

$\sum x = 5{,}500 \qquad \sum x^2 = 3{,}036{,}600 \qquad \sum xy = 182{,}045{,}000$

$\sum y = 331{,}500 \qquad \sum y^2 = 11{,}155{,}250{,}000$

$$SS_{xy} = \sum xy - \frac{\sum x \sum y}{n} = 182{,}045{,}000 - \frac{5{,}500(331{,}500)}{10} = -280{,}000$$

$$SS_{xx} = \sum x^2 - \frac{(\sum x)^2}{n} = 3{,}036{,}600 - \frac{5{,}500^2}{10} = 11{,}600$$

$$SS_{yy} = \sum y^2 - \frac{(\sum y)^2}{n} = 11{,}155{,}250{,}000 - \frac{5{,}500(331{,}500)}{10} = 166{,}025{,}000$$

$$\hat{\beta}_1 = \frac{SS_{xy}}{SS_{xx}} = \frac{-280{,}000}{11{,}600} = -24.13793103 \approx -24.14$$

$$\hat{\beta}_0 = \bar{y} - \hat{\beta}_1 \bar{x} = \frac{331{,}500}{10} - (-24.13793103)\frac{5{,}500}{10} = 46{,}425.86207 \approx 46{,}426$$

The estimated regression line is $\hat{y} = 46{,}426 - 24.14x$.

$$\text{SSE} = \text{SS}_{yy} - \hat{\beta}_1\text{SS}_{xy} = 166{,}025{,}000 - (-24.13793103)(-280{,}000) = 159{,}266{,}379.3$$

$$s^2 = \frac{\text{SSE}}{n - 2} = \frac{159{,}266{,}379.3}{10 - 2} = 19{,}908{,}297.41$$

$$\bar{x} = \frac{5{,}500}{10} = 550$$

We want to predict the GMAT score, x, of an MBA student with a starting salary of $y_p = \$35{,}000$. The estimate of x is calculated as follows:

$$\hat{x} = \frac{y_p - \hat{\beta}_0}{\hat{\beta}_1} = \frac{35{,}000 - 46{,}426}{-24.14} = 473.32$$

An appropriate 95% prediction interval for x is $\hat{x} \pm t_{.025}\left(\frac{s}{\hat{\beta}_1}\right)\sqrt{1 + \frac{1}{n} + \frac{(\hat{x} - \bar{x})^2}{\text{SS}_{xx}}}$

where $t_{.025} = 2.360$ is based on 8 df.

Substitution yields:

$$473.32 \pm 2.306\left(\frac{\sqrt{19{,}908{,}297.41}}{-24.14}\right)\sqrt{1 + \frac{1}{10} + \frac{(473.32 - 550)^2}{11{,}600}}$$

$$\Rightarrow 473.32 \pm 540.3 \Rightarrow (-66.98, 1013.62)$$

b. You can see that the interval contains nonsensical values for GMAT score, x. This is due to the fact that the straight-line model is inadequate for predicting starting salary, y. Therefore, the model is also inadequate for inverse prediction.

8.9 a. The results of the simple linear regression are given below:

$n = 8$	$\text{SS}_{xx} = 7.219$	$\hat{\beta}_1 = 6.06$	
$\bar{x} = 2.563$	$\text{SS}_{xy} = 43.75$	$\hat{\beta}_0 = -2.03$	
$\bar{y} = 13.50$	$\text{SS}_{yy} = 280.00$	$\text{SSE} = 14.85$	$s^2 = 2.47$

The least squares prediction equation is $\hat{y} = -2.03 + 6.06x$.

b. H_0: $\beta_1 = 0$
H_a: $\beta_1 \neq 0$

The test statistic: $t = \dfrac{\hat{\beta}_1}{s/\sqrt{\text{SS}_{xx}}} = \dfrac{6.06}{\sqrt{2.47/7.219}} = 10.35$

Rejection region: $\alpha = .05$, df $= n - 2 = 6$, $t_{.025} = 2.447$
Reject H_0 if $t < -2.447$ or $t > 2.447$

Conclusion: Reject H_0 at $\alpha = .05$. There is sufficient evidence to indicate that the model is useful for predicting decrease in pulse rate y.

c. First, we calculate the estimate of x for $y_p = 10$:

$$\hat{x} = \frac{y_p - \hat{\beta}_0}{\hat{\beta}_1} = \frac{10 - (-2.03)}{6.06} = 1.985$$

An appropriate 95% prediction interval for x is:

$$\hat{x} \pm t_{.025}(6 \text{ df})\left(\frac{s}{\hat{\beta}_1}\right)\sqrt{1 + \frac{1}{n} + \frac{(\hat{x} - \bar{x})^2}{SS_{xx}}}$$

$$\Rightarrow 1.985 \pm 2.447\left(\frac{\sqrt{2.47}}{6.06}\right)\sqrt{1 + \frac{1}{8} + \frac{(1.985 - 2.563)^2}{7.219}} \Rightarrow 1.985 \pm .687$$

or (1.298, 2.672)

Therefore, to reduce a patient's pulse rate to $y = 10$ beats/minute, we should administer a dosage x of the drug somewhere between 1.298 cc and 2.672 cc.

8.11 a. The following information is obtained from the SAS regression printout:

$\hat{y} = -1.2667 + .1760x$

$r^2 = .5629$, MSE $= 46.2564$, $s = 6.801$, $s_{\hat{\beta}_1} = .043$

$t(\text{for } H_0\text{: } \beta_1 = 0) = 4.09$, p-value $= .0013$

Since the p-value (.0013) associated with the test statistic for testing H_0: $\beta_1 = 0$, is smaller than $\alpha = .05$, we have sufficient evidence to reject H_0 and conclude that the model is useful for predicting number of defectives y.

b. The residuals, calculated in the table, are plotted against x in the following figure.

Observation i	1	2	3	4	5	6	7	8
x_i	100.00	100.00	100.00	100.00	100.00	150.00	150.00	150.00
y_i	15.00	23.00	11.00	14.00	18.00	19.00	29.00	20.00
$\hat{y}_i$	16.33	16.33	16.33	16.33	16.33	25.13	25.13	25.13
Residual = $(y_i - \hat{y}_i)$	−1.33	6.67	−5.33	−2.33	1.67	−6.13	3.87	−5.13

Observation i	9	10	11	12	13	14	15
x_i	150.00	150.00	200.00	200.00	200.00	200.00	200.00
y_i	35.00	24.00	26.00	48.00	27.00	38.00	30.00
$\hat{y}_i$	25.13	25.13	33.93	33.93	33.93	33.93	33.93
Residual = $(y_i - \hat{y}_i)$	9.87	−1.13	−7.93	14.07	−6.93	4.07	−3.93

The plot reveals a funnel shape, that is, as machine speed x increases, the spread of the residuals also increases. This implies that the assumption of equal variances is very likely violated.

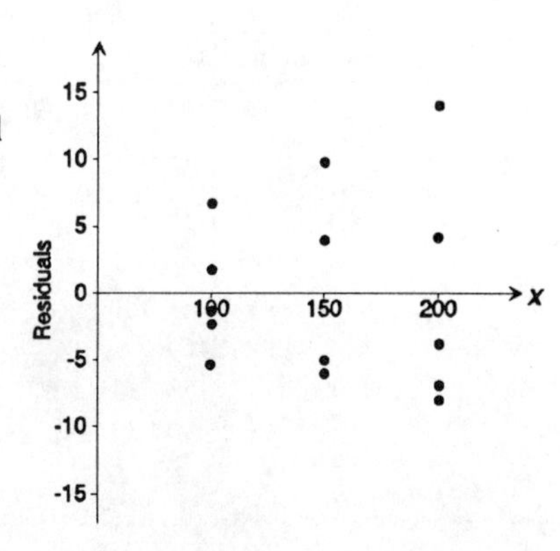

c. Let $\hat{\epsilon}_i = (y_i - \hat{y}_i)$ represent the residual for the ith-observation. To obtain the variance of the 5 residuals at each level of x, we use the short-cut formula for the sample variance.

$$s^2 = \frac{\sum_{i=1}^{5} \hat{\epsilon}_i^2 - \frac{\left[\sum_{i=1}^{5} \hat{\epsilon}_i\right]^2}{5}}{5 - 1}$$

The results are shown in the following table:

x	Variance of Residuals (s^2)	s^2/x	s^2/x^2	$s^2/\sqrt{x}$
100	20.7	.2070	.00207	2.070
150	44.3	.2953	.00197	3.617
200	85.2	.4260	.00213	6.025

Note that when the variance s^2 is divided by x^2 for each level of x, the result is a constant (approximately .002). Therefore, the variance is proportional to $1/x^2$, and the appropriate weights to use in a weighted least squares are:

$$w_i = \frac{1}{x_i^2}$$

d. The weighted least squares fit, obtained from the SAS regression procedure, is:

$$\hat{y} = -1.51428571 + .17771429x$$

The standard deviation of the estimate of the weighted least squares slope, $s_{\hat{\beta}_1}$ is .040, a slight decrease from the unweighted value of .043.

e. The weighted residuals are defined as $\sqrt{w_i}\,(y_i - \hat{y}_i)$.

Since $w_i = \frac{1}{x_i^2}$, we have $\sqrt{w_i}\,(y_i - \hat{y}_i) = \sqrt{\frac{1}{x_i^2}}(y_i - \hat{y}_i) = (y_i - \hat{y}_i)/x_i$

The weighted residuals, calculated in the table below, are plotted against x_i in the figure.

Observation i	1	2	3	4	5	6	7	8
x_i	100.00	100.00	100.00	100.00	100.00	150.00	150.00	150.00
y_i	15.00	23.00	11.00	14.00	18.00	19.00	29.00	20.00
$\hat{y}_i$	16.26	16.26	16.26	16.26	16.26	25.14	25.14	25.14
$(y_i - \hat{y}_i)$	−1.26	6.74	−5.26	−2.26	1.74	−6.14	3.86	−5.14
$(y_i - \hat{y}_i)/x_i$	−.013	.067	−.053	−.023	.017	−.041	.026	−.034

Observation i	9	10	11	12	13	14	15
x_i	150.00	150.00	200.00	200.00	200.00	200.00	200.00
y_i	35.00	24.00	26.00	48.00	27.00	38.00	30.00
$\hat{y}_i$	25.14	25.14	34.03	34.03	34.03	34.03	34.03
$(y_i - \hat{y}_i)$	9.86	−1.14	−8.03	13.97	−7.03	3.97	−4.03
$(y_i - \hat{y}_i)/x_i$	.066	.008	−.040	.070	−.035	.020	−.020

From the residual plot, it is evident that the spread of the residuals for the three values of x is constant. Thus, it appears that weighted least squares has corrected the problem of unequal variances.

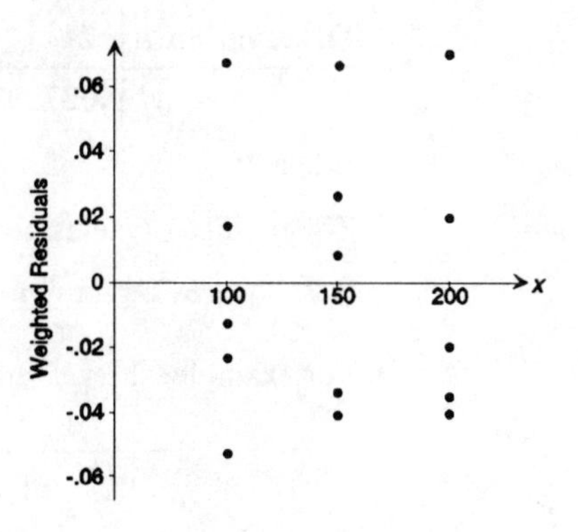

8.13 When the response (i.e., dependent) variable y is recorded as 0 or 1, three major problems arise:

1. The assumption of normal errors is violated.
2. The assumption of equal error variances is violated.
3. Predicted values may take on nonsensical values (i.e., negative values or values greater than 1).

The first two problems are the most serious, since we have little or no faith in any inferences derived from an ordinary least squares regression model when the standard regression assumptions are violated.

8.15 a. β_1 = Slope of the $y - x_1$ line, holding x_2 and x_3 constant [i.e., the mean increase (or decrease) in the probability of being hired for each additional year of education, for applicants of the same sex and experience].

β_2 = Slope of the $y - x_2$ line, holding x_1 and x_3 constant [i.e., the mean increase (or decrease) in the probability of being hired for each additional year of experience, for applicants of the same sex and education].

$\beta_3 = E(y)_{x_3=1} - E(y)_{x_3=0}$, holding x_1 and x_2 constant

[i.e., the mean difference between the probability of males and the probability of females being hired, for applicants of the same education and experience].

Note: β_0 has no practical interpretation in this model.

b. **Stage 1:**

The ordinary least squares fit, obtained from the SAS regression package, is:

$$\hat{y} = -.844356 + .106919x_1 + .088634x_2 + .484728x_3$$

The predicted values $\hat{y}_i$ for the 28 observations are listed below:

Observation i	1	2	3	4	5	6	7	8	9	10
$\hat{y}_i$	−.023	.068	.814	.548	−.328	.277	.245	−.062	−.114	.897

Observation i	11	12	13	14	15	16	17	18	19	20
$\hat{y}_i$	.245	.454	−.239	.418	.511	.152	.496	.371	.204	.157

Observation i	21	22	23	24	25	26	27	28
$\hat{y}_i$	.027	.282	.939	.381	.100	.371	.954	.8615

Stage 2:

The predicted values above are used to calculated the appropriate weights for weighted least squares regression, where $w_i = \dfrac{1}{\hat{y}_i(1 - \hat{y}_i)}$.

For example, the weight associated with observation #15 is:

$$w_{15} = \frac{1}{\hat{y}_{15}(1 - \hat{y}_{15})} = \frac{1}{(.511)(.489)} = 4.002$$

The weighted least squares fit (obtained using SAS) is:

$$\hat{y} = -0.527915 + .074979x_1 + .074750x_2 + .391172x_3$$

c. H_0: $\beta_1 = \beta_2 = \beta_3 = 0$
H_a: At least one $\beta \neq 0$

Test statistic: $F = 21.79$ (p-value $= .0001$), obtained from the SAS printout.

Rejection region: Using the p-value approach, reject H_0 if p-value $< \alpha = .05$.

Conclusion: Reject H_0 at $\alpha = .05$. There is sufficient evidence to indicate the model is useful for predicting hiring status, y.

d. H_0: $\beta_3 = 0$
H_a: $\beta_3 \neq 0$

Test statistic: $t = 4.01$ (p-value $= .0005$), obtained from the SAS printout

Rejection region: Using the p-value approach, reject H_0 if p-value $< \alpha = .05$.

Conclusion: Reject H_0 at $\alpha = .05$. There is sufficient evidence to indicate that sex is an important predictor of hiring status y. Since the estimate of β_3 is positive, it appears that males have a higher probability of being hired than females.

e. A 95% confidence interval for $E(y)$ when $x_1 = 4$, $x_2 = 3$, and $x_3 = 0$ (obtained from the SAS printout) is (−.2122, .2047). That is, we estimate the probability of being hired $E(y)$, for female applicants ($x_3 = 0$) with $x_1 = 4$ years of education and $x_2 = 3$ years of experience to fall between −.2122 and .2047. Note that the interval includes negative numbers, which are nonsensical probabilities.

8.17 a. H_0: $\beta_1 = \beta_2 = \beta_3 = 0$
H_a: At least one coefficient is not 0

Test statistic: $X^2 = 20.430$ (from printout); p-value $= .0001$

Rejection region: Using the p-value approach, reject H_0 if the p-value is less than $\alpha = .05$.

Conclusion: Reject H_0 at $\alpha = .05$. There is sufficient evidence to indicate the model is adequate.

b. H_0: $\beta_3 = 0$
H_a: $\beta_3 \neq 0$

Test statistic: $X^2 = 4.6352$
p-value $= .0313$

Rejection region: Using the p-value approach, reject H_0 if the p-value is less than $\alpha = .05$.

Conclusion: Reject H_0 at $\alpha = .05$. There is sufficient evidence to indicate the sex is an important predictor of hiring status.

c. Observation number 11 has $x_1 = 4$, $x_2 = 0$, and $x_3 = 1$. The 95% confidence interval for the mean response is (.00048, .40027). We are 95% confident that the probability that a male applicant with 4 years of higher education and 0 years of experience is hired is between .00048 and .40027.

9 Time Series Modeling and Forecasting

9.1 a. The quarterly times series is shown here.

It appears that the long term trend in housing starts decreases for a period and then increases.

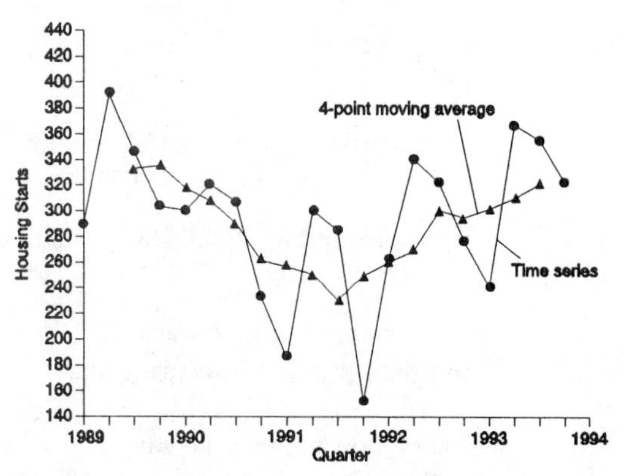

b. To compute a 4-point moving average, first find the sum of the time series values from two periods before, one period before, the current period, and one period ahead. The sum is $S_t = Y_{t-2} + Y_{t-1} + Y_t + Y_{t+1}$. The moving average is found by dividing the sum by 4.

For 1989,III: $S_{1989,III} = Y_{1989,I} + Y_{1989,II} + Y_{1989,III} + Y_{1989,IV}$
$= 290.6 + 390.9 + 346.2 + 303.6 = 1331.3$

For 1989,IV: $S_{1989,IV} = Y_{1989,II} + Y_{1989,III} + Y_{1989,IV} + Y_{1990,I}$
$= 390.9 + 346.2 + 303.6 + 300.8 = 1341.5$

The rest of the sums are found in a similar manner and are found in the table.

To compute the moving averages, the sums are divided by 4.

For 1989,III: $M_{1989,III} = \frac{S_{1989,III}}{4} = \frac{1331.3}{4} = 332.83$

For 1989,IV: $M_{1989,IV} = \frac{S_{1989,IV}}{4} = \frac{1341.5}{4} = 335.38$

The rest of the moving average values are found in a similar manner and are found in the table.

Year	Quarter	Housing Starts	4-point Moving Total	4-point Moving Average
1989	I	290.6		
	II	390.9		
	III	346.2	1331.3	332.83
	IV	303.6	1341.5	335.38
1990	I	300.8	1271.4	317.85
	II	320.8	1232.3	308.08
	III	307.1	1161.7	290.43
	IV	233.0	1046.3	261.58
1991	I	185.4	1026.3	256.58
	II	300.8	1004.0	251.00
	III	284.8	923.0	230.75
	IV	152.0	999.6	249.90
1992	I	262.0	1039.4	259.85
	II	340.6	1076.7	269.18
	III	322.1	1201.4	300.35
	IV	276.7	1180.0	295.00
1993	I	240.6	1206.6	301.65
	II	367.2	1240.1	310.03
	III	355.6	1285.4	321.35
	IV	322.0		

c. By graphing the moving average on the graph in part (a), we can see a decreasing and then increasing trend better. The moving average has smoothed out the seasonal effect in the time series.

d. To find the seasonal index, we first find the ratio of y_t divided by the corresponding moving average, M_t. The seasonal index is then the average of these ratios for a particular quarter multiplied by 100. For quarter I, the ratios are shown in the table:

Year	Quarter	y_t	M_t	y_t/M_t
1989	I	290.6	–	–
1990	I	300.8	317.85	.946
1991	I	185.4	256.58	.723
1992	I	262.0	259.85	1.008
1993	I	240.6	301.65	.798

The seasonal index for quarter I is:

$$\frac{.946 + .723 + 1.008 + .798}{4}(100) = 86.875$$

e. For quarter II, the ratios are shown in the table:

Year	Quarter	y_t	M_t	y_t/M_t
1989	II	390.9	–	–
1990	II	320.8	308.08	1.041
1991	II	300.8	251.00	1.198
1992	II	340.6	269.18	1.265
1993	I	367.2	310.03	1.184

The seasonal index for quarter II is:

$$\frac{1.041 + 1.198 + 1.265 + 1.184}{4}(100) = 117.2$$

f. To forecast using the moving average, we extend the graph of the moving average two time periods.

$F_{1994,\text{I}} \approx 330$ and $F_{1994,\text{II}} \approx 340$

Using the seasonal index, the new forecasts are:

$$F_{1994,\text{I}} = 330\frac{\text{Seasonal Index}}{100} = 330(86.875/100) = 286.69$$

$$F_{1994,\text{I}} = 340\frac{\text{Seasonal Index}}{100} = 340(117.2/100) = 398.48$$

9.3 a. Refer to Exercise 9.2 for the computation of the exponentially smoothed series and the Holt-Winter series. These series are repeated below:

Year	Quarter	Housing Starts	Exp	E_t	T_t	S_t
1989	I	290.6	290.60			
	II	390.9	310.66	390.90	110.30	1.00
	III	346.2	317.77	462.20	85.80	0.75
	IV	303.6	314.93	499.12	61.36	0.61
1990	I	300.8	312.11	508.54	35.39	0.59
	II	320.8	313.85	499.31	13.08	0.64
	III	307.1	312.50	491.91	2.84	0.66
	IV	233.0	296.60	472.41	−8.33	0.53
1991	I	185.4	274.36	433.95	−23.39	0.48
	II	300.8	279.65	422.08	−17.63	0.69
	III	284.8	280.68	409.64	−15.04	0.69
	IV	152.0	254.94	373.29	−25.70	0.44
1992	I	262.0	256.35	388.04	−5.47	0.62
	II	340.6	273.20	404.55	5.52	0.80
	III	322.1	282.98	422.07	11.52	0.74
	IV	276.7	281.73	471.70	30.57	0.54
1993	I	240.6	273.50	479.99	19.43	0.54
	II	367.2	292.24	491.70	15.57	0.76
	III	355.6	304.91	501.96	12.91	0.72
	IV	322.0	308.33	530.36	20.66	0.59

From Exercise 9.1, the forecasts for 1994,I and 1994,II using the moving average are:

$$F_{1994,I} = 286.69$$
$$F_{1994,II} = 398.48$$

The forecasts using the exponentially smoothed series are:

$$F_t = wy_n + (1 - w)E_n, \text{ for } t = n + 1, n + 2, \ldots$$

Thus: $F_{1994,I} = .2(322.0) + .8(308.33) = 311.06$
$F_{1994,II} = .2(322.0) + .8(308.33) = 311.06$

The forecasts using the Holt-Winters series are:

$$F_t = (E_n + kT_n)S_{n+1-P}, \text{ for } t = n + k; k = 1, 2, \ldots ; (P = 4)$$

Thus: $F_{1994,I} = (530.36 + 20.66)(.54) = 297.55$
$F_{1994,II} = (530.36 + 2[20.66])(.76) = 434.48$

To check for accuracy of the overall forecasts, we will use the MAD measure.

$$\text{MAD} = \frac{\sum |F_t - y_t|}{m}$$

Moving Average:

$$\text{MAD} = \frac{|286.69 - 293.9| + |398.48 - 421.6|}{2} = 15.17$$

Exponential Smoothing:

$$\text{MAD} = \frac{|311.06 - 293.9| + |311.06 - 421.6|}{2} = 63.85$$

Holt-Winters:

$$\text{MAD} = \frac{|297.55 - 293.9| + |433.48 - 421.6|}{2} = 8.27$$

Thus, using the MAD measure, the Holt-Winters forecast is the most accurate.

b. To check for accuracy of the overall forecasts, we will use the RMSE measure.

$$\text{RMSE} = \sqrt{\frac{\sum (F_t - y_t)^2}{m}}$$

Moving Average:

$$\text{RMSE} = \sqrt{\frac{(286.69 - 293.9)^2 + (398.48 - 421.6)^2}{2}} = 17.12$$

Exponential Smoothing:

$$\text{RMSE} = \sqrt{\frac{(311.06 - 293.9)^2 + (311.06 - 421.6)^2}{2}} = 79.10$$

Holt-Winters:

$$\text{RMSE} = \sqrt{\frac{(297.55 - 293.9)^2 + (434.48 - 421.6)^2}{2}} = 9.47$$

Again, using the RMSE measure, the Holt-Winters forecast is the most accurate.

c. Using both the MAD and the RMSE measure, the Holt-Winters forecast is the most accurate.

9.5 a. The graph of the time series is shown here.

The CPI increases over time.

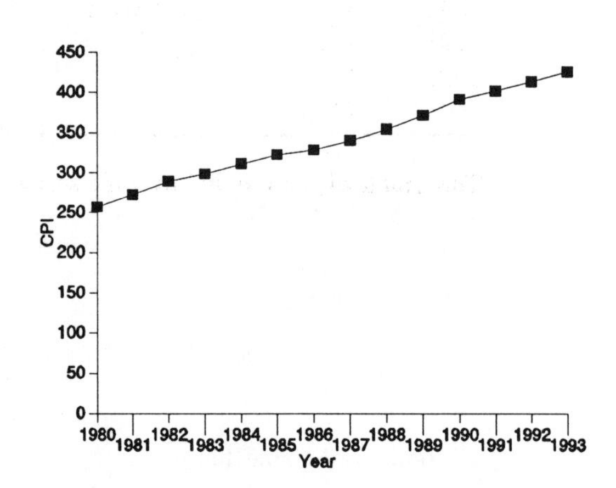

b. To compute a 5-point moving average, first find the sum of the time series values from two periods before, one period before, the current period, one period ahead, and two periods ahead. The sum is $S_t = Y_{t-2} + Y_{t-1} + Y_t + Y_{t+1} + Y_{t+2}$. The moving average is found by dividing the sum by 5.

For 1982, $S_{1982} = Y_{1980} + Y_{1981} + Y_{1982} + Y_{1983} + Y_{1984}$
$= 246.8 + 272.4 + 289.1 + 298.4 + 311.1 = 1417.8$

For 1983, $S_{1983} = Y_{1981} + Y_{1982} + Y_{1983} + Y_{1984} + Y_{1985}$
$= 272.4 + 289.1 + 298.4 + 311.1 + 322.2 = 1493.2$

The rest of the sums are found in a similar manner and are shown in the table below.

To compute the moving averages, the sums are divided by 5.

For 1982, $M_{1982} = \frac{S_{1982}}{5} = \frac{1417.8}{5} = 283.56$

For 1983, $M_{1983} = \frac{S_{1983}}{5} = \frac{1493.2}{5} = 298.64$

The rest of the moving average values are found in a similar manner and are shown in the table below.

Year	CPI	5-point Moving Total	5-point Moving Average
1980	246.8		
1981	272.4		
1982	289.1	1417.8	283.56
1983	298.4	1493.2	298.64
1984	311.1	1549.2	309.84
1985	322.2	1600.2	320.04
1986	328.4	1656.0	331.20
1987	340.1	1716.2	343.24
1988	354.2	1785.3	357.06
1989	371.3	1858.5	371.70
1990	391.3	1931.6	386.32
1991	401.6	2003.0	400.60
1992	413.2		
1993	425.6		

The graph of the moving average series is:

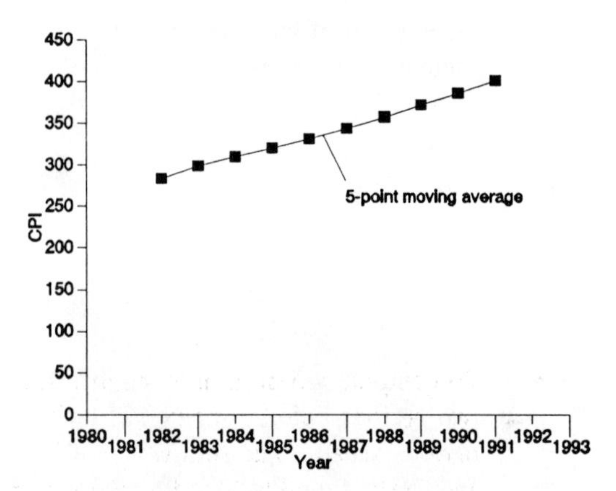

The time period 1995 is four steps from the last computed moving average value. To forecast this value, we need to extend the graph of the moving average to 1995. Doing this, we get a value of approximately 456. Thus,

$$F_{1995} = 456$$

c. To compute the exponentially smoothed series, we use the following:

$$E_1 = Y_1$$
$$E_t = wY_t + (1 - w)E_{t-1} \text{ for } t > 1$$

Thus, the exponentially smoothed series is:

$$E_1 = 246.8$$
$$E_2 = .4(272.4) + (1 - .4)(246.8) = 257.04$$
$$E_3 = .4(289.1) + (1 - .4)(257.04) = 269.86$$

The rest of the values of the exponentially smoothed series are found in the same manner and are shown in the following table:

Year	CPI	E_t
1980	246.8	246.80
1981	272.4	257.04
1982	289.1	269.86
1983	298.4	281.28
1984	311.1	293.21
1985	322.2	304.80
1986	328.4	314.24
1987	340.1	324.59
1988	354.2	336.43
1989	371.3	350.38
1990	391.3	366.75
1991	401.6	380.69
1992	413.2	393.69
1993	425.6	406.46

The graph of the exponentially smoothed series is:

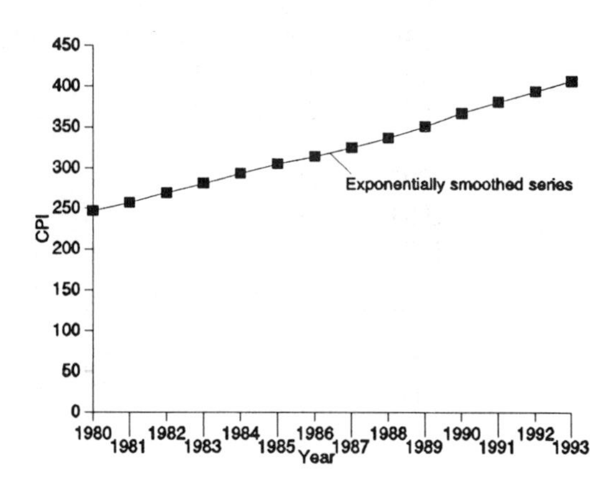

The time period 1995 is two steps from the last observed time series value. To forecast this value, we need to use the following:

$$F_t = wY_n + (1 - w)E_n$$

Thus: $F_{1995} = .4(425.6) + (1 - .4)(406.46) = 414.12$

d. To compute the Holt-Winters series with trend only, we use the following:

$$E_t = \begin{cases} y_2, \ t = 2 \\ wy_t + (1 - w)(E_{t-1} + T_{t-1}), \ t > 2 \end{cases}$$

$$T_t = \begin{cases} y_2 - y_1, \ t = 2 \\ v(E_t - E_{t-1}) + (1 - v)T_{t-1}, \ t > 2 \end{cases}$$

Thus, the Holt-Winters series is:

$E_2 = 272.4$
$E_3 = .4(289.1) + (1 - .4)(272.4 + 25.6) = 294.44$
$E_4 = .4(298.4) + (1 - .4)(294.44 + 23.82) = 304.32$

$T_2 = 272.4 - 246.8 = 25.6$
$T_3 = .5(294.44 - 272.4) + (1 - .5)25.6 = 23.82$
$T_4 = .5(304.32 - 294.44) + (1 - .5)23.82 = 16.85$

The rest of the values of the Holt-Winters series are found in a similar manner and are shown in the following table:

Year	CPI	E_t	T_t
1980	246.8		
1981	272.4	272.40	25.60
1982	289.1	294.44	23.82
1983	298.4	310.32	19.85
1984	311.1	322.54	16.04
1985	322.2	332.02	12.76
1986	328.4	338.23	9.48
1987	340.1	344.67	7.96
1988	354.2	353.26	8.27
1989	371.3	365.44	10.23
1990	391.3	381.92	13.35
1991	401.6	397.81	14.62
1992	413.2	412.73	14.77
1993	425.6	426.75	14.39

The time period 1995 is two steps from the last observed time series value. To forecast this value, we need to use the following:

$$F_t = E_n + kT_n, \text{ for } t = k + n, k = 1, 2, \ldots$$

Thus, $F_{1995} = 426.75 + 2(14.39) = 455.53$

9.7 a. To compute a 3-point moving average, first find the sum of the time series values from one period before, the current period, and one period ahead. The sum is $S_t = Y_{t-1} + Y_t + Y_{t+1}$. The moving average is found by dividing the sum by 3.

For 1972, $S_{1972} = Y_{1971} + Y_{1972} + Y_{1973} = 41.25 + 58.61 + 97.81 = 197.67$
For 1973, $S_{1973} = Y_{1972} + Y_{1973} + Y_{1974} = 58.61 + 97.81 + 159.70 = 316.12$

The rest of the sums are found in a similar manner and are shown in the table that follows.

To compute the moving averages, the sums are divided by 3.

For 1972, $M_{1972} = \frac{S_{1972}}{3} = \frac{197.67}{3} = 65.896$

For 1973, $M_{1973} = \frac{S_{1973}}{3} = \frac{316.12}{3} = 105.37$

The rest of the moving average values are found in a similar manner and are shown in the table below.

Year	Price of Gold	3-Point Moving Total	3-Point Moving Average
1971	41.25		
1972	58.61	197.67	65.89
1973	97.81	316.12	105.37
1974	159.70	418.91	139.64
1975	161.40	445.90	148.63
1976	124.80	434.50	144.83
1977	148.30	466.60	155.53
1978	193.50	649.60	216.53
1979	307.80	1107.31	369.10
1980	606.01	1364.44	454.81
1981	450.63	1430.82	476.94
1982	374.18	1273.84	424.61
1983	449.03	1183.50	394.50
1984	360.29	1126.62	375.54
1985	317.30	1045.46	348.49
1986	367.87	1094.08	364.69
1987	408.91	1213.71	404.57
1988	436.93	1227.05	409.02
1989	381.21	1202.21	400.74
1990	384.07		

The graph of the time series and the 3-point moving average is:

The long-term trend is increasing. However, there is no evidence of a cyclical pattern.

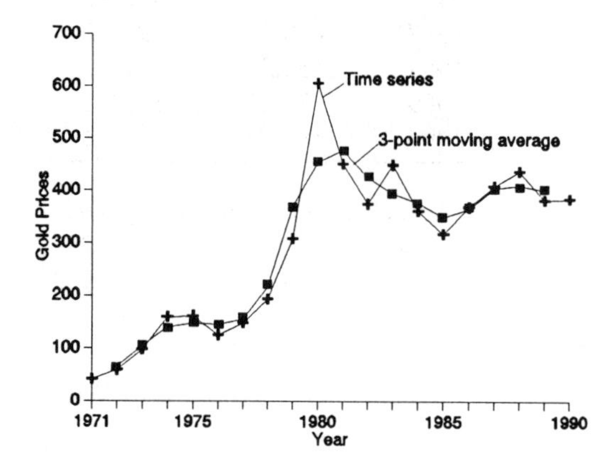

b. To forecast the price of gold for 1991, 1992, and 1993, we need to extend the graph of the moving average to 1993. Doing this, we get the following:

$$F_{1991} \approx 400; F_{1992} \approx 400; F_{1993} \approx 400$$

c. To compute the exponentially smoothed series, we use the following:

$$E_1 = Y_1$$
$$E_t = wY_t + (1 - w)E_{t-1} \text{ for } t > 1$$

Thus, the exponentially smoothed series is:

$$E_1 = 41.25$$
$$E_2 = .8(58.61) + (1 - .8)(41.25) = 55.14$$
$$E_3 = .8(97.81) + (1 - .8)(55.14) = 89.28$$

The rest of the values of the exponentially smoothed series are found in the same manner and are shown in the following table:

Year	Price of Gold	E_t
1971	41.25	41.25
1972	58.62	55.14
1973	97.81	89.28
1974	159.70	145.62
1975	161.40	158.24
1976	124.80	131.49
1977	148.30	144.94
1978	193.50	183.79
1979	307.80	283.00
1980	606.01	541.41
1981	450.63	468.79
1982	374.18	393.10
1983	449.03	437.84
1984	360.29	375.80
1985	317.30	329.00
1986	367.87	360.10
1987	408.91	399.15
1988	436.93	429.37
1989	381.21	390.84
1990	384.07	385.42

The plot of the exponentially smoothed series is:

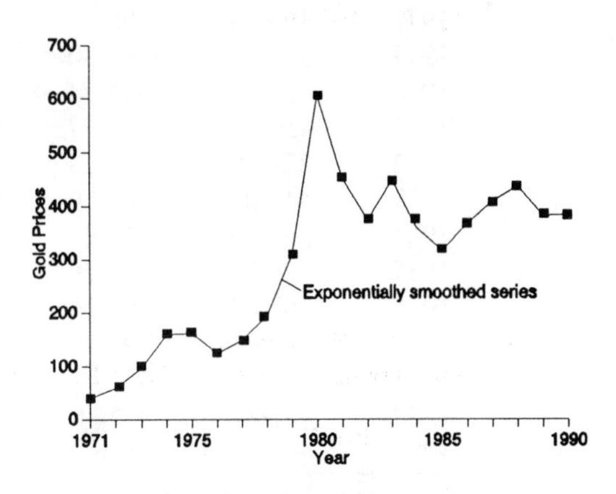

d. To forecast using the exponentially smoothed series, we need to use the following:

$$F_t = wY_n + (1 - w)E_n, \text{ for } t = n + 1, n + 2, \ldots$$

Thus:

$$F_{1991} = .8(384.07) + (1 - .8)(385.42) = 384.34$$
$$F_{1992} = .8(384.07) + (1 - .8)(385.42) = 384.34$$
$$F_{1993} = .8(384.07) + (1 - .8)(385.42) = 384.34$$

e. To compute the Holt-Winters series with trend only, we use the following:

$$E_t = \begin{cases} y_2, \ t = 2 \\ wy_t + (1 - w)(E_{t-1} + T_{t-1}), \ t > 2 \end{cases}$$

$$T_t = \begin{cases} y_2 - y_1, \ t = 2 \\ v(E_t - E_{t-1}) + (1 - v)T_{t-1} \ t > 2 \end{cases}$$

Thus, the Holt-Winters series is:

$$E_2 = 58.61$$
$$E_3 = .8(97.81) + (1 - .8)(58.61 + 17.36) = 93.44$$
$$E_4 = .8(159.7) + (1 - .8)(93.44 + 24.35) = 151.32$$

$$T_2 = 58.61 - 41.25 = 17.36$$
$$T_3 = .4(93.44 - 58.61) + (1 - .4)17.36 = 24.35$$
$$T_4 = .4(151.32 - 93.44) + (1 - .4)24.35 = 37.76$$

The rest of the values of the Holt-Winters series are found in a similar manner and are shown in the following table:

Year	Price of Gold	E_t	T_t
1971	41.25		
1972	58.62	58.61	17.36
1973	97.81	93.44	24.35
1974	159.70	151.32	37.76
1975	161.40	166.94	28.90
1976	124.80	139.01	6.17
1977	148.30	147.68	7.17
1978	193.50	185.77	19.54
1979	307.80	287.30	52.34
1980	606.01	552.74	137.58
1981	450.63	498.57	60.88
1982	374.18	411.23	1.59
1983	449.03	441.79	13.18
1984	360.29	379.23	−17.12
1985	317.30	326.26	−31.46
1986	367.87	353.26	−8.08
1987	408.91	396.16	12.32
1988	436.93	431.24	21.42
1989	381.21	395.50	−1.44
1990	384.07	386.07	−4.64

To forecast using the Holt-Winters series, we need to use the following:

$$F_t = E_n + kT_n, \text{ for } t = k + n, k = 1, 2, \ldots$$

Thus:

$$F_{1991} = 386.07 + (-4.64) = 381.43$$
$$F_{1992} = 386.07 + 2(-4.64) = 376.79$$
$$F_{1993} = 386.07 + 3(-4.64) = 372.15$$

f. We will use the MAD and RMSE measures to assess the accuracy of the three forecasting methods. First, we will compute the MAD measure for each method.

$$\text{MAD} = \frac{\sum |F_t - y_t|}{m}$$

Moving Average:

$$\text{MAD} = \frac{|400 - 362.04| + |400 - 344.50| + |400 - 383.69|}{3} = 36.59$$

Exponential Smoothing:

$$\text{MAD} = \frac{|384.34 - 362.04| + |384.34 - 344.50| + |384.34 - 383.69|}{3}$$
$$= 20.93$$

Holt-Winters:

$$\text{MAD} = \frac{|381.43 - 362.04| + |376.79 - 344.50| + |372.15 - 383.69|}{3}$$
$$= 21.07$$

Thus, using the MAD measure, the exponential smoothing method and the Holt-Winters forecast are almost identical in accuracy.

To check for accuracy of the overall forecasts, we will use the RMSE measure.

$$\text{RMSE} = \sqrt{\frac{\sum (F_t - y_t)^2}{m}}$$

Moving Average:

$$\text{RMSE} = \sqrt{\frac{(400 - 362.04)^2 + (400 - 344.50)^2 + (400 - 383.69)^2}{3}} = 39.95$$

Exponential Smoothing:

$$\text{RMSE} = \sqrt{\frac{(384.34 - 362.04)^2 + (384.34 - 344.50)^2 + (384.34 - 383.69)^2}{3}}$$
$$= 26.36$$

Holt-Winters:

$$\text{RMSE} = \sqrt{\frac{(381.43 - 362.04)^2 + (376.79 - 344.50)^2 + (372.15 - 383.69)^2}{3}}$$
$$= 22.74$$

Using the RMSE measure, the Holt-Winters forecast is the most accurate. Thus, overall, the Holt-Winters series appears to be the most accurate.

9.9 a. The graph of the weekly series y_t, shown at right reveals strong visual evidence of a quadratic trend.

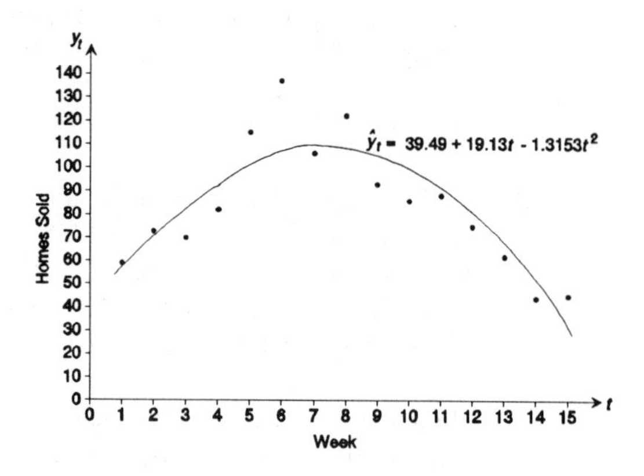

b. The printout using SAS is:

DEP VARIABLE: Y

ANALYSIS OF VARIANCE

SOURCE	DF	SUM OF SQUARES	MEAN SQUARE	F VALUE	PROB>F
MODEL	2	8162.80795	4081.40398	21.599	0.0001
ERROR	12	2267.59205	188.96600		
C TOTAL	14	10430.40000			

ROOT MSE	13.74649	R-SQUARE	0.7826
DEP MEAN	83.8	ADJ R-SQ	0.7464
C.V.	16.40393		

PARAMETER ESTIMATES

VARIABLE	DF	PARAMETER ESTIMATE	STANDARD ERROR	T FOR H0: PARAMETER=0	PROB > |T|
INTERCEP	1	39.48791209	12.24446298	3.225	0.0073
T	1	19.13031674	3.52154451	5.432	0.0002
TSQ	1	-1.31528765	0.21402392	-6.146	0.0001

OBS	ACTUAL	PREDICT VALUE	STD ERR PREDICT	LOWER95% PREDICT	UPPER95% PREDICT	RESIDUAL
1	59.0000	57.3029	9.3709	21.0546	93.5512	1.6971
2	73.0000	72.4874	7.1173	38.7599	106.2	0.5126
3	70.0000	85.0413	5.5953	52.7042	117.4	-15.0413
4	82.0000	94.9646	4.8705	63.1891	126.7	-12.9646
5	115.0	102.3	4.7908	70.5394	134.0	12.7427
6	137.0	106.9	5.0150	75.0374	138.8	30.0805
7	106.0	109.0	5.2506	76.8895	141.0	-2.9510
8	122.0	108.4	5.3440	76.2173	140.5	13.6480
9	93.0000	105.1	5.2506	73.0609	137.2	-12.1225
10	86.0000	99.2623	5.0150	67.3803	131.1	-13.2623
11	88.0000	90.7716	4.7908	59.0537	122.5	-2.7716
12	75.0000	79.6503	4.8705	47.8749	111.4	-4.6503
13	62.0000	65.8984	5.5953	33.5613	98.2355	-3.8984
14	44.0000	49.5160	7.1173	15.7885	83.2434	-5.5160
15	45.0000	30.5029	9.3709	-5.7454	66.7512	14.4971
16	.	8.8593	12.2445	-31.2506	48.9693	.

SUM OF RESIDUALS	5.79092E-13
SUM OF SQUARED RESIDUALS	2267.592
PREDICTED RESID SS (PRESS)	3525.982

The least squares prediction equation is:

$$\hat{y}_t = 39.48791209 + 19.13031674t - 1.31528765t^2$$

c. The least squares prediction equation $\hat{y}_t$, plotted in the graph above, appears to provide an adequate fit to the quadratic secular trend. The coefficient of determination is $R^2 = .7826$ and the test statistic for checking model adequacy is $F = 21.60$ (p-value $= .0001$).

d. A 95% prediction interval for home sales in week 16(y_{16}), obtained from the SAS printout, is (−31.2506, 48.9693). Note the negative lower bound and the large width of the interval. Although the quadratic model is statistically adequate (see part (c)), the wide 95% prediction interval reveals that the model may not be very practical for forecasting weekly home sales.

9.11 a. In general, it appears that each mode of transportation has a secular trend that is adequately represented by this model.

b. The printouts for fitting the model to each of these data sets are:

```
RAILROADS:

DEP VARIABLE: Y1
                         ANALYSIS OF VARIANCE

                      SUM OF          MEAN
SOURCE     DF        SQUARES        SQUARE       F VALUE        PROB>F

MODEL       1     5600.72803    5600.72803        71.550        0.0001
ERROR       8      626.22097   78.27762121
C TOTAL     9     6226.94900

    ROOT MSE        8.847464      R-SQUARE        0.8994
    DEP MEAN           29.21      ADJ R-SQ        0.8869
    C.V.            30.28916

                         PARAMETER ESTIMATES

                   PARAMETER          STANDARD        T FOR H0:
VARIABLE  DF        ESTIMATE             ERROR     PARAMETER=0  PROB > |T|

INTERCEP   1     74.52666667        6.04396861          12.331      0.0001
T          1     -8.23939394        0.97407374          -8.459      0.0001

                          PREDICT    STD ERR   LOWER95%    UPPER95%
OBS          ACTUAL         VALUE    PREDICT    PREDICT     PREDICT  RESIDUAL

      1     67.1000       66.2873     5.2001    42.6217     89.9529    0.8127
      2     74.3000       58.0479     4.4103    35.2510     80.8447   16.2521
      3     46.3000       49.8085     3.7092    27.6856     71.9314   -3.5085
      4     36.5000       41.5691     3.1564    19.9071     63.2310   -5.0691
      5     28.6000       33.3297     2.8399    11.9019     54.7575   -4.7297
      6     17.9000       25.0903     2.8399     3.6625     46.5181   -7.1903
      7      7.3000       16.8509     3.1564    -4.8110     38.5129   -9.5509
      8      5.8000        8.6115     3.7092   -13.5114     30.7344   -2.8115
      9      4.7000        0.3721     4.4103   -22.4247     23.1690    4.3279
     10      3.6000       -7.8673     5.2001   -31.5329     15.7983   11.4673
     11           .      -24.3461      6.922   -50.2507      1.5586         .

SUM OF RESIDUALS              2.42473E-13
SUM OF SQUARED RESIDUALS          626.221
PREDICTED RESID SS (PRESS)       1084.633
```

BUSES:

DEP VARIABLE: Y2

ANALYSIS OF VARIANCE

SOURCE	DF	SUM OF SQUARES	MEAN SQUARE	F VALUE	PROB>F
MODEL	1	491.90427	491.90427	13.134	0.0067
ERROR	8	299.61673	37.45209091		
C TOTAL	9	791.52100			

ROOT MSE	6.119811	R-SQUARE	0.6215
DEP MEAN	21.83	ADJ R-SQ	0.5742
C.V.	28.03395		

PARAMETER ESTIMATES

VARIABLE	DF	PARAMETER ESTIMATE	STANDARD ERROR	T FOR H0: PARAMETER=0	PROB > ¦T¦
INTERCEP	1	35.26000000	4.18062704	8.434	0.0001
T	1	-2.44181818	0.67376905	-3.624	0.0067

OBS	ACTUAL	PREDICT VALUE	STD ERR PREDICT	LOWER95% PREDICT	UPPER95% PREDICT	RESIDUAL
1	26.5000	32.8182	3.5969	16.4486	49.1877	-6.3182
2	21.4000	30.3764	3.0506	14.6077	46.1450	-8.9764
3	37.7000	27.9345	2.5656	12.6321	43.2370	9.7655
4	32.4000	25.4927	2.1833	10.5091	40.4763	6.9073
5	25.7000	23.0509	1.9644	8.2293	37.8725	2.6491
6	24.2000	20.6091	1.9644	5.7875	35.4307	3.5909
7	16.9000	18.1673	2.1833	3.1837	33.1509	-1.2673
8	14.2000	15.7255	2.5656	0.4230	31.0279	-1.5255
9	11.4000	13.2836	3.0506	-2.4850	29.0523	-1.8836
10	7.9000	10.8418	3.5969	-5.5277	27.2114	-2.9418
11	.	5.9582	4.788	-11.9601	23.8765	.

SUM OF RESIDUALS	8.85958E-14
SUM OF SQUARED RESIDUALS	299.6167
PREDICTED RESID SS (PRESS)	495.6202

AIR CARRIERS:

DEP VARIABLE: Y3

ANALYSIS OF VARIANCE

SOURCE	DF	SUM OF SQUARES	MEAN SQUARE	F VALUE	PROB>F
MODEL	1	9829.09394	9829.09394	295.268	0.0001
ERROR	8	266.31006	33.28875758		
C TOTAL	9	10095.40400			

ROOT MSE	5.769641	R-SQUARE	0.9736
DEP MEAN	46.86	ADJ R-SQ	0.9703
C.V.	12.31251		

PARAMETER ESTIMATES

VARIABLE	DF	PARAMETER ESTIMATE	STANDARD ERROR	T FOR H0: PARAMETER=0	PROB > ¦T¦
INTERCEP	1	-13.17333333	3.94141517	-3.342	0.0102
T	1	10.91515152	0.63521657	17.183	0.0001

OBS	ACTUAL	PREDICT VALUE	STD ERR PREDICT	LOWER95% PREDICT	UPPER95% PREDICT	RESIDUAL
1	2.8000	-2.2582	3.3911	-17.6911	13.1747	5.0582
2	2.7000	8.6570	2.8761	-6.2094	23.5233	-5.9570
3	14.3000	19.5721	2.4188	5.1453	33.9990	-5.2721
4	28.9000	30.4873	2.0583	16.3610	44.6135	-1.5873
5	42.1000	41.4024	1.8520	27.4289	55.3760	0.6976
6	54.7000	52.3176	1.8520	38.3440	66.2911	2.3824
7	73.1000	63.2327	2.0583	49.1065	77.3590	9.8673
8	77.7000	74.1479	2.4188	59.7210	88.5747	3.5521
9	83.9000	85.0630	2.8761	70.1967	99.9294	-1.1630
10	88.4000	95.9782	3.3911	80.5453	111.4	-7.5782
11	.	117.8	4.514	100.9	134.7	.

SUM OF RESIDUALS	3.08642E-14
SUM OF SQUARED RESIDUALS	266.3101
PREDICTED RESID SS (PRESS)	457.2737

The least squares prediction equations are:

Railroads: $\hat{y}_t = 74.527 - 8.239t$
Buses: $\hat{y}_t = 35.260 - 2.442t$
Air Carriers: $\hat{y}_t = -13.173 + 10.915t$

c. The plots with the least squares prediction equation are:

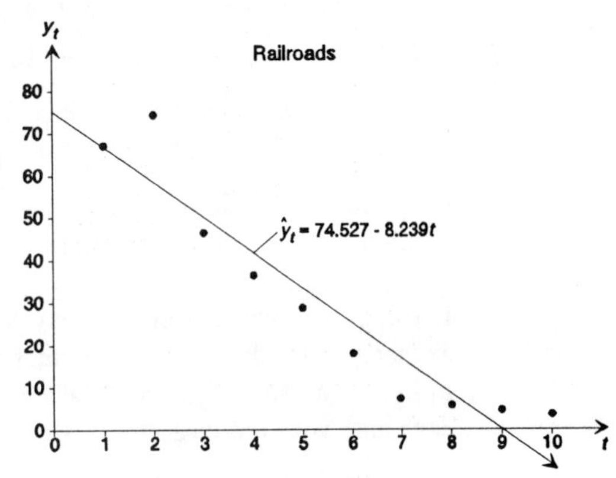

The least squares prediction equation appears to provide an adequate fit to the secular trend. The coefficient of determination is $R^2 = .8994$, and the test statistic for checking model adequacy is $F = 71.55$. ($p-$value $= .0001$)

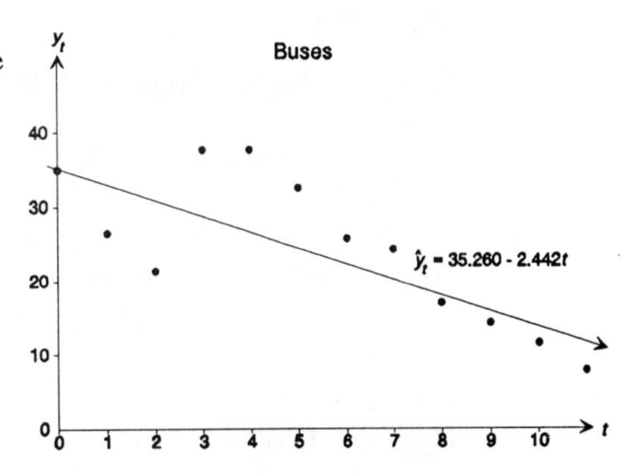

The least squares prediction equation appears to provide an adequate fit to the secular trend. The coefficient of determination is $R^2 = .6215$, and the test statistic for checking model adequacy is $F = 13.134$. ($p-$value $= .0067$)

The least squares prediction equation appears to provide an adequate fit to the secular trend. The coefficient of determination is $R^2 = .9736$, and the test statistic for checking model adequacy is $F = 295.268$. (p-value $= .0001$)

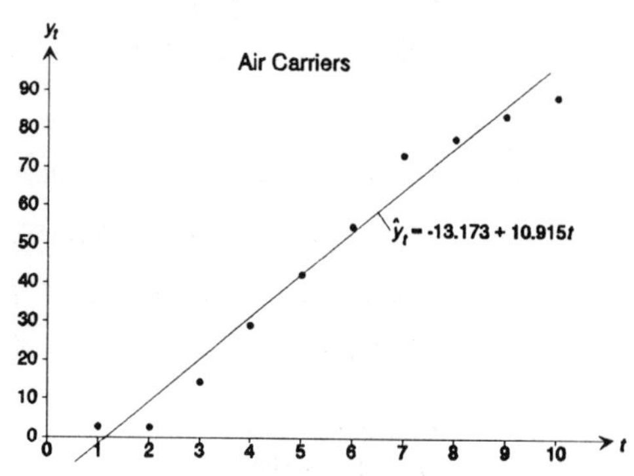

d. Railroads: $\hat{y}_t = 74.527 - 8.239(12) = -24.35$

Using the SAS regression procedure, the 95% prediction interval is $(-50.26, 1.57)$ (11th observation)

Buses: $\hat{y}_t = 35/26 - 2.442(12) = 5.96$

Using the SAS regression procedure, the 95% prediction interval is $(-11.96, 23.88)$ (11th observation)

Air Carriers: $\hat{y}_t = -13.173 + 10.915(12) = 117.81$

Using the SAS regression procedure, the 95% prediction interval is $(100.91, 134.71)$ (11th observation)

In all three of these predictions, we are predicting outside the observed range. The relationship may not be the same outside the observed range.

9.13 a. Let μ_{PRE} = mean difference between the April rates of return of the two stocks on the exchange with the largest and smallest returns in a year in the pre-tax period and let μ_{POST} = mean difference between the April rates of return of the two stocks on the exchange with the largest and smallest returns in a year in the post-tax period.

Then, $\beta_1 = \mu_{\text{POST}} - \mu_{\text{PRE}}$

b. $\beta_0 = \mu_{\text{PRE}}$

c. During the pre-tax period, $D_t = 0$. Thus, $\hat{y}_t = -.55 + 3.08(0) = -.55$

d. During the post-tax period, $D_t = 1$. Thus, $\hat{y}_t = -.55 + 3.08(1) = 2.53$

e. In 1995, the value of D_t is 1 (post-tax period). Thus, a forecast for the difference in April rates of return of the two stocks in 1995 is:

$$\hat{y}_{1995} = -.55 + 3.08(1) = 2.53$$

9.15 a. H_0: $\beta_2 = 0$
H_a: $\beta_2 < 0$

Test statistic: $t = -1.39$

Rejection region: For $\alpha = .05$ and df $= n - 2 = 105$, $t_{.05} = 1.645$;
Reject H_0 if $t < -1.645$.

Conclusion: Do not reject H_0 at $\alpha = .05$. There is insufficient evidence to indicate that quarterly number of pension plan qualifications y_t increases at a decreasing rate over time, t.

b. Substituting $t = 108$ into the prediction equation, we obtain:

$$\hat{y}_{108} = 6.19 + .039(108) - .00024(108)^2 = 7.603$$

Since y_t is the natural log of the number of pension plan qualifications, we must take the antilog of $\hat{y}_{108}$ to obtain the forecast: Forecast $= e^{\hat{y}_{108}} = e^{7.603} = 2003.48$

c. H_0: $\beta_2 = 0$
H_a: $\beta_2 < 0$

Test statistic: $t = -1.61$

Rejection region: For $\alpha = .05$ and df $= n - 2 = 105$, $t_{.05} = 1.645$; Reject H_0 if $t < -1.645$.

Conclusion: Do not reject H_0 at $\alpha = .05$. There is insufficient evidence to indicate that quarterly number of profit sharing plan qualifications y_t increases at a decreasing rate over time, t.

d. Substituting $t = 108$ into the prediction equation, we obtain:

$$\hat{y}_{108} = 6.22 + .035(108) - .00021(108)^2 = 7.551$$

The forecast is the antilog of $\hat{y}_{108}$: Forecast $= e^{\hat{y}_{108}} = e^{7.551} = 1901.81$

9.17 a. The nonzero autocorrelations are those for values of $m = 4, 8, 12, 16$, and 20:

Autocorrelation $(R_t, R_{t+4}) = \phi^{4/4} = \phi^1 = .5$
Autocorrelation $(R_t, R_{t+8}) = \phi^{8/4} = \phi^2 = .25$
Autocorrelation $(R_t, R_{t+12}) = \phi^{12/4} = \phi^3 = .125$
Autocorrelation $(R_t, R_{t+16}) = \phi^{16/4} = \phi^4 = .0625$
Autocorrelation $(R_t, R_{t+20}) = \phi^{20/4} = \phi^5 = .03125$

Thus, the first twenty autocorrelations for this model are:

0, 0, 0, .5, 0, 0, 0, .25, 0, 0, 0, .125, 0, 0, 0, .0625, 0, 0, 0, .03125

b.

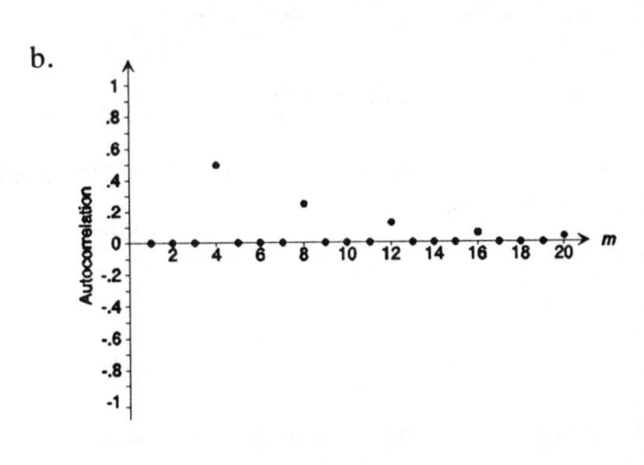

The first twenty autocorrelations for the first-order model $R_t = .5R_{t-1} + \epsilon_t$ are plotted below for comparison.

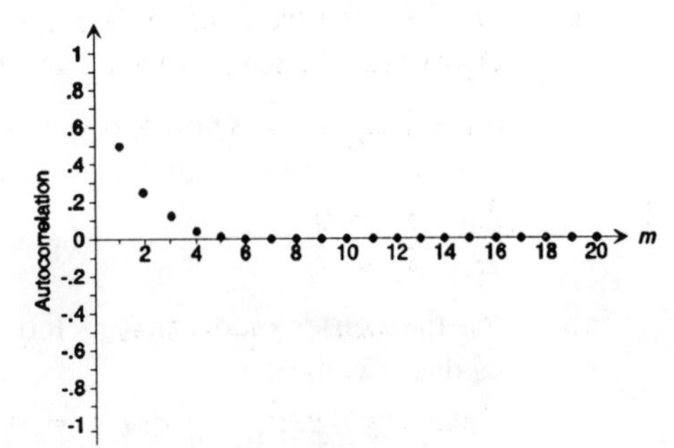

9.19 The general form of the pth-order autoregressive model is:

$$R_t = \phi_1 R_{t-1} + \phi_2 R_{t-2} + \cdots + \phi_p R_{t-p} + \epsilon_t$$

Thus, for $p = 4$, we have: $R_t = \phi_1 R_{t-1} + \phi_2 R_{t-2} + \phi_3 R_{t-3} + \phi_4 R_{t-4} + \epsilon_t$

9.21 a. The first-order model is: $E(y_t) = \beta_0 + \beta_1 x_{1t} + \beta_2 x_{2t} + \beta_3 x_{3t} + \beta_4 t$

b. The interaction model allows for interaction between each independent variable and time:

$$E(y_t) = \beta_0 + \beta_1 x_{1t} + \beta_2 x_{2t} + \beta_3 x_{3t} + \beta_4 t + \beta_5 x_{1t} t + \beta_6 x_{2t} t + \beta_7 x_{3t} t$$

c. We postulate a first-order autoregressive model for R_t:

$$R_t = \phi R_{t-1} + \epsilon_t,$$

where $|\phi| < 1$ and ϵ_t is a white noise process. This model is appropriate because we expect that residuals at consecutive time points will be positively correlated (i.e., they will tend to have the same sign), and the correlation will diminish as the time distance between two points increases.

9.23 a. The model for daily price with mean $E(y_t)$ may be written as:

$$E(y_t) = \beta_0 + \beta_1 \left[\cos \frac{2\pi t}{365}\right] + \beta_2 \left[\sin \frac{2\pi t}{365}\right]$$

This cyclic model, with a period of 365 days, allows for seasonality, with the expected peaks and valleys remaining the same from year to year.

b. The amplitude (magnitude of the seasonal effect) and phase shift (which determines the location in time of the peaks and valleys) are functions of β_1 and β_2. The value of β_0 is the overall mean of the time series.

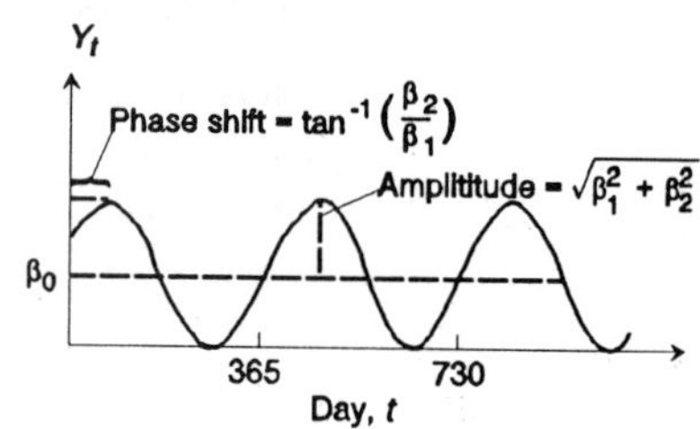

c. The interaction model is:

$$E(y_t) = \beta_0 + \beta_1 \left[\cos \frac{2\pi t}{365}\right] + \beta_2 \left[\sin \frac{2\pi t}{365}\right] + \beta_3 t + \beta_4 t \left[\cos \frac{2\pi t}{365}\right] + \beta_5 t \left[\sin \frac{2\pi t}{365}\right]$$

The presence of interaction indicates seasonal effects increase or decrease with time (not the same from year to year).

d. We expect that residuals on consecutive days would be positively correlated. Thus, it is unreasonable to assume that the random error component, R_t, is white noise. Instead, we propose a first-order autoregressive model for R_t:

$$R_t = \phi R_{t-1} + \epsilon_t, \quad \text{where } |\phi| < 1 \text{ and } \epsilon_t \text{ is a white noise process.}$$

9.25 a. The time series model that includes a straight-line long-term trend and autocorrelated residuals would be:

$$y_t = \beta_0 + \beta_1 t + \phi R_{t-1} + \epsilon_t$$

b. Using the SAS AUTOREG output, the least squares prediction equation is:

$$\hat{y}_t = 4787.82 + 22.34t + .736\hat{R}_{t-1}$$

c. $\hat{\beta}_0 = 4787.82$: This is the estimate of where the regression line will cross the y_t axis.

$\hat{\beta}_1 = 22.34$: The estimated increase in mean GDP for each additional time period is 22.34.

$\hat{\phi} = .736$: Since $\hat{\phi}$ is positive, it implies that the time series residuals are positively autocorrelated.

d. $R^2 = .8991$. This means that 89.91% of the sum of squares of the deviation of the GDP values about their means is attributable to the time series model.

ROOT MSE = 43.607. We would expect most of the observations to fall within $2s$ or 2(43.607) or 87.214 units of their predicted values.

9.27 a. $\hat{\beta}_0 = 434.93$: This is an estimate of where the regression line will cross the y_t axis.

$\hat{\beta}_1 = 92.280$: The estimated increase in mean DJA for each additional year is 92.280.

$\hat{\phi} = .721$: Since $\hat{\phi}$ is positive, it implies that the time series residuals are positively correlated.

b. $R^2 = .9051$: 90.51% of the variability in the DJA values is explained by the autoregressive model.

9.29 a. We begin the forecasting procedure by calculating the estimated residual for the last observation in the series, i.e., $\hat{R}_n$, where

$$\hat{R}_n = y_n - \hat{y}_n = y_n - (\hat{\beta}_0 + \hat{\beta}_1 n + \hat{\beta}_2 n^2)$$

Substituting $n = 48$, $\hat{\beta}_0 = 220$, $\hat{\beta}_1 = 17$, $\hat{\beta}_2 = -.3$, and $y_{48} = 350$, we have:

$$\begin{aligned}\hat{R}_{48} &= y_{48} - [\hat{\beta}_0 + \hat{\beta}_1(48) + \hat{\beta}_2(48)^2] \\ &= 350 - [220 + 17(48) - .3(48)^2] \\ &= 350 - 334.8 = 5.2\end{aligned}$$

To obtain the forecasts, we use the formulas:

$$\hat{R}_t = .81\hat{R}_{t-1}$$
$$\hat{y}_t = 220 + 17t - .3t^2 + \hat{R}_t$$

$t = 49$: $\hat{R}_{49} = .81\hat{R}_{48} = .81(5.2) = 4.212$

$\hat{y}_{49} = 220 + 17(49) - .3(49)^2 + 4.212 = 336.91$

$t = 50$: $\hat{R}_{50} = .81\hat{R}_{49} = .81(4.212) = 3.412$

$\hat{y}_{50} = 220 + 17(50) - .3(50)^2 + 3.412 = 323.41$

$t = 51$: $\hat{R}_{51} = .81\hat{R}_{50} = .81(3.412) = 2.763$

$\hat{y}_{51} = 220 + 17(51) - .3(51)^2 + 2.763 = 309.46$

b. An approximate 95% prediction interval for the forecast m time periods into the future is given by the formula:

$$\hat{y}_{n+m} \pm 2\sqrt{\text{MSE}(1 + \hat{\phi}^2 + \hat{\phi}^4 + \cdots + \hat{\phi}^{2(m-1)})}$$

Since $n = 48$, we want to obtain prediction intervals for the forecasts $m = 1$, $m = 2$, and $m = 3$ time periods (quarters) into the future.

Approximate 95% Prediction Intervals

$t = 49$: $\hat{y}_{49} \pm 2\sqrt{\text{MSE}}$

$(m = 1) \Rightarrow 336.91 \pm 2\sqrt{10.5} \Rightarrow 336.91 \pm 6.48 \Rightarrow (330.43, 343.39)$

$t = 50$: $\hat{y}_{50} \pm 2\sqrt{\text{MSE}(1 + \hat{\phi}^2)}$

$(m = 2) \Rightarrow 323.41 \pm 2\sqrt{10.5(1 + .81^2)} \Rightarrow 323.41 \pm 8.34 \Rightarrow (315.07, 331.75)$

$t = 51$: $\hat{y}_{51} \pm 2\sqrt{\text{MSE}(1 + \hat{\phi}^2 + \hat{\phi}^4)}$

$(m = 3) \Rightarrow 309.46 \pm 2\sqrt{10.5(1 + .81^2 + .81^4)} \Rightarrow 309.46 \pm 9.36 \Rightarrow (300.10, 318.82)$

9.31 For quarter I, 1994, $t = 17$:

$$\hat{R}_{16} = Y_{16} - \hat{Y}_{16} = 5218 - (4787.82 + 22.34(16)) = 72.74$$

$$\hat{Y}_{17} = 4787.82 + 22.34t + .736\hat{R}_{16} = 4787.82 + 22.34(17) + .736(72.74) = 5221.14$$

The form of the prediction interval is:

$$\hat{Y}_{n+m} \pm 2\sqrt{\text{MSE}(1 + \hat{\phi}^2 + \hat{\phi}^4 + \cdots + \hat{\phi}^{2(m-1)})}$$

where m is the number of steps ahead.

For $t = 17$, $m = 1$. The approximate 95% prediction interval is:

$$\hat{Y}_{17} \pm 2\sqrt{\text{MSE}} \Rightarrow 5221.14 \pm 2(43.607) \Rightarrow 5221.14 \pm 87.21 \Rightarrow (5133.93, 5308.35)$$

The observed value for quarter I, 1994 was 5,261. This number falls in the prediction interval.

For quarter II, 1994, $t = 18$:

$$\hat{R}_{17} = .736(\hat{R}_{16}) = .736(72.74) = 53.537$$

$$\hat{Y}_{18} = 4787.82 + 22.34t + .736\hat{R}_{17} = 4787.82 + 22.34(18) + .736(53.537) = 5229.34$$

The form of the prediction interval is:

$$\hat{Y}_{n+m} \pm 2\sqrt{\text{MSE}(1 + \hat{\phi}^2 + \hat{\phi}^4 + \cdots + \hat{\phi}^{2(m-1)})}$$

where m is the number of steps ahead.

For $t = 18$, $m = 2$. The approximate 95% prediction interval is:

$$\hat{Y}_{18} \pm 2\sqrt{\text{MSE}(1 + \hat{\phi}^2)} \Rightarrow 5229.34 \pm 2(43.607)\sqrt{1 + .736^2}$$
$$\Rightarrow 5229.34 \pm 108.29 \Rightarrow (5121.05, 5337.63)$$

The observed value for quarter II, 1994 was 5,314. This number falls in the prediction interval.

For quarter III, 1994, $t = 19$:

$$\hat{R}_{18} = .736(\hat{R}_{17}) = .736(53.537) = 39.403$$
$$\hat{Y}_{19} = 4787.82 + 22.34t + .736\hat{R}_{18} = 4787.82 + 22.34(19) + .736(39.403) = 5241.28$$

The form of the prediction interval is:

$$\hat{Y}_{n+m} \pm 2\sqrt{\text{MSE}(1 + \hat{\phi}^2 + \hat{\phi}^4 + \cdots + \hat{\phi}^{2(m-1)})}$$

where m is the number of steps ahead.

For $t = 19$, $m = 3$. The approximate 95% prediction interval is:

$$\hat{Y}_{19} \pm 2\sqrt{\text{MSE}(1 + \hat{\phi}^2 + \phi^4)} \Rightarrow 5241.28 \pm 2(43.607)\sqrt{1 + .736^2 + .736^4}$$
$$\Rightarrow 5241.28 \pm 118.15 \Rightarrow (5123.13, 5359.43)$$

The observed value for quarter III, 1994 was 5,359. This number falls in the prediction interval.

9.33 For 1991, $t = 21$:

$$\hat{R}_{20} = Y_{20} - \hat{Y}_{20} = 2679 - (434.93 + 92.28(20)) = 398.47$$
$$\hat{Y}_{21} = 434.93 + 92.28t + .721\hat{R}_{20} = 434.93 + 92.28(21) + .721(398.47) = 2,660.11$$

The form of the prediction interval is:

$$\hat{Y}_{n+m} \pm 2\sqrt{\text{MSE}(1 + \hat{\phi}^2 + \hat{\phi}^4 + \cdots + \hat{\phi}^{2(m-1)})}$$

where m is the number of steps ahead.

For $t = 21$, $m = 1$. The approximate 95% prediction interval is:

$$\hat{Y}_{21} \pm 2\sqrt{\text{MSE}} \Rightarrow 2,660.11 \pm 2(203.43) \Rightarrow 2,660.11 \pm 406.86 \Rightarrow (2,253.25, 3,066.97)$$

The observed value for 1991 was 3,015. This number falls in the prediction interval.

For 1992, $t = 22$:

$$\hat{R}_{21} = .721(\hat{R}_{20}) = .721(398.47) = 287.30$$
$$\hat{Y}_{22} = 434.93 + 92.28t + .721\hat{R}_{21} = 434.93 + 92.28(22) + .721(287.30) = 2,672.23$$

The form of the prediction interval is:

$$\hat{Y}_{n+m} \pm 2\sqrt{\text{MSE}(1 + \hat{\phi}^2 + \hat{\phi}^4 + \cdots + \hat{\phi}^{2(m-1)})}$$

where m is the number of steps ahead.

For $t = 22$, $m = 2$. The approximate 95% prediction interval is:

$$\hat{Y}_{22} \pm 2\sqrt{\text{MSE}(1 + \hat{\phi}^2)} \Rightarrow 2,672.23 \pm 2(203.43)\sqrt{1 + .721^2}$$
$$\Rightarrow 2,672.23 \pm 504.05 \Rightarrow (2,168,18, 3,176.28)$$

The observed value for 1992 was 3,160. This number falls in the prediction interval.

For 1993, $t = 23$:

$$\hat{R}_{22} = .721(\hat{R}_{21}) = .721(287.30) = 207.14$$
$$\hat{Y}_{23} = 434.93 + 92.28t + .721\hat{R}_{22} = 434.93 + 92.28(23) + .721(207.14) = 2{,}706.72$$

The form of the prediction interval is:

$$\hat{Y}_{n+m} \pm 2\sqrt{\text{MSE}\left(1 + \hat{\phi}^2 + \hat{\phi}^4 + \cdots + \hat{\phi}^{2(m-1)}\right)}$$

where m is the number of steps ahead.

For $t = 23$, $m = 3$. The approximate 95% prediction interval is:

$$\hat{Y}_{23} \pm 2\sqrt{\text{MSE}(1 + \hat{\phi}^2 + \hat{\phi}^4)} \Rightarrow 2{,}706.72 \pm 2(203.43)\sqrt{1 + .721^2 + .721^4}$$
$$\Rightarrow 2{,}706.72 \pm 544.35 \Rightarrow (2{,}162{,}37,\ 3{,}251.07)$$

The observed value for 1993 was 3,520. This number does not fall in the prediction interval.

9.35 a. Recall that $\hat{y}_t = 3.54 + .039t + \hat{R}_t$, where $\hat{R}_t = .04\hat{R}_{t-1}$.

To forecast y_{108}, we first must obtain the estimate of the residual for the last (107th) quarter, $\hat{R}_{107}$:

$$\hat{R}_{107} = y_{107} - \hat{y}_{107} = y_{107} - [\hat{\beta}_0 + \hat{\beta}_1(107)]$$

Substituting $y_{107} = 7.5$, $\hat{\beta}_0 = 3.54$ and $\hat{\beta}_1 = .039$, we have:

$$\hat{R}_{107} = 7.5 - [3.54 + .039(107)] = 7.5 - 7.713 = -.213$$

The forecast for y_{108} is calculated as follows: $\hat{y}_{108} = 3.54 + .039(108) + \hat{R}_{108}$

where $\hat{R}_{108} = .40\hat{R}_{107} = .40(-.213) = -.0852$

Substituting, we have $\hat{y}_{108} = 3.54 + .039(108) - .0852 = 7.6668$

Finally, to obtain the forecast of number of pension plan qualifications in quarter $t = 108$, we take the antilog:

$$\text{Forecasted number of pension plan qualifications} = e^{\hat{y}_{108}} = e^{7.668} = 2{,}136.23$$

b. Since we are forecasting only $m = 1$ time period (quarter) into the future, an approximate 95% prediction interval for y_{108} is:

$$\hat{y}_{108} \pm 2\sqrt{\text{MSE}} \Rightarrow 7.6668 \pm 2\sqrt{.0440} \Rightarrow 7.6668 \pm .4195 \text{ or } (7.2473,\ 8.0863)$$

Approximate 95% confidence bounds for the forecast of quarterly number of pension plan qualifications are obtained by taking antilogs of the endpoints of the interval for y_{108}:

$$(e^{7.2473}, e^{8.0863}) = (1404.28,\ 3249.72)$$

Therefore, we forecast number of pension plan qualifications in quarter 4 of 1982 (i.e., $t = 108$) to fall between 1404.28 and 3249.72 with 95% confidence.

c. Recall that $\hat{y}_t = 3.45 + .038t + \hat{R}_t$, where $\hat{R}_t = .22\hat{R}_{t-1}$. First we calculate:

$$\begin{aligned}\hat{R}_{107} &= y_{107} - \hat{y}_{107} = y_{107} - [\hat{\beta}_0 + \hat{\beta}_1(107)] \\ &= 7.6 - [3.45 + .308(107)] = 7.6 - 7.516 = .084\end{aligned}$$

Then $\hat{R}_{108} = .22\hat{R}_{107} = .22(.084) = .0185$

and $\hat{y}_{108} = 3.45 + .308(108) + \hat{R}_{108} = 3.45 + .038(108) + .0185 = 7.5725$

Taking the antilog, our forecast of number of profit-sharing plan terminations in quarter $t = 108$ is $e^{7.5727} = 1943.96$.

An approximate 95% prediction interval for y_{108} is:

$$\hat{y}_{108} \pm 2\sqrt{\text{MSE}} \Rightarrow 7.5725 \pm 2\sqrt{.0402} \Rightarrow 7.5725 \pm .4010 \Rightarrow (7.1715, 7.9735)$$

Taking antilogs, we obtain 95% confidence bounds for the number of profit-sharing plan terminations in quarter $t = 108$:

$$(e^{7.1715}, e^{7.9735}) = (1301.80, 2903.00)$$

9.37 a. To compute a 12-point moving average, first find the sum of the time series values from six, five, four, three, two, and one periods before, the current period, one, two, three, four, and five periods ahead. The sum is $S_t = Y_{t-6} + Y_{t-5} + \cdots + Y_t + Y_{t+1} + \cdots + Y_{t+5}$. The moving average is found by dividing the sum by 12.

For Jul., 1991, $S_{J,1991} = Y_{J,1991} + Y_{F,1991} + Y_{M,1991} + \cdots + Y_{D,1991}$
$= 8.0 + 6.7 + 8.2 + \cdots + 6.5 = 98$

For Aug., 1991, $S_{A,1991} = Y_{F,1991} + Y_{M,1991} + Y_{A,1991} + \cdots + Y_{J,1992}$
$= 6.7 + 8.2 + 9.4 + \cdots + 7.2 = 97.2$

The rest of the sums are found in a similar manner and are shown in the table that follows.

To compute the moving averages, the sums are divided by 12.

For Jul., 1991, $M_{J,1991} = \dfrac{S_{J,1991}}{12} = \dfrac{98}{12} = 8.17$

For Aug., 1991, $M_{A,1991} = \dfrac{S_{a,1991}}{12} = \dfrac{97.2}{12} = 8.10$

The rest of the moving average values are found in a similar manner and are shown in the table below.

Year	Month	Mortgage Appl.	Moving Total	Moving Average
1991	Jan.	8.0		
	Feb.	6.7		
	Mar.	8.2		
	Apr.	9.4		
	May	10.5		
	June	8.8		
	July	10.1	98.0	8.17
	Aug.	7.4	97.2	8.10
	Sep.	7.1	98.0	8.17
	Oct.	8.7	99.9	8.33
	Nov.	6.6	99.9	8.33
	Dec.	6.5	97.3	8.11
1992	Jan.	7.2	96.2	8.02
	Feb.	7.5	94.9	7.91
	Mar.	10.1	95.0	7.92
	Apr.	9.4	95.0	7.92
	May	7.9	93.6	7.80
	June	7.7	94.1	7.84
	July	8.8	94.1	7.84
	Aug.	7.5	92.6	7.72
	Sep.	7.1	92.1	7.68
	Oct.	7.3	89.7	7.48
	Nov.	7.1	88.5	7.38
	Dec.	6.5	88.3	7.36
1993	Jan.	5.7	89.0	7.42
	Feb.	7.0	88.5	7.38
	Mar.	7.7	89.6	7.47
	Apr.	8.2	89.9	7.49
	May	7.7	91.4	7.62
	June	8.4	93.5	7.79
	July	8.3	94.1	7.84
	Aug.	8.6		
	Sep.	7.4		
	Oct.	8.8		
	Nov.	9.2		
	Dec.	7.1		

The plot of the time series and the 12-point moving average is:

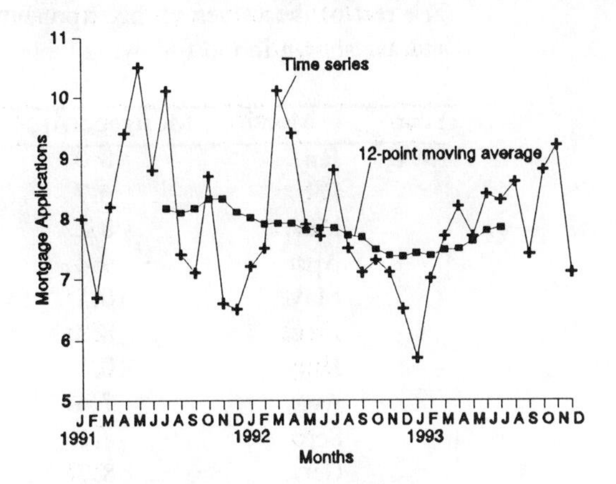

The series tends to decrease and then increase over the long run. This is the secular trend. Also, there does appears to be a seasonal pattern. The series tends to be high in April through June and low from December to January.

b. Because there appears to be a seasonal pattern, we must first compute the seasonal index for January. To find the seasonal index, we first find the ratio of y_t divided by the corresponding moving average M_t. The seasonal index is then the average of these ratios for a particular month multiplied by 100. For January, the ratios are shown in the table:

Year	Month	y_t	M_t	y_t/M_t
1992	Jan.	7.2	8.02	.898
1993	Jan.	5.7	7.42	.768

The seasonal index for January is:

$$\frac{.898 + .768}{2}(100) = 83.3$$

To forecast using the moving average, we extend the graph of the moving average through January of 1994.

$$F_{J,1994} \approx 8.4$$

Using the seasonal index, the new forecasts are:

$$F_{J,1994} = 8.4\frac{\text{Seasonal Index}}{100} = 8.4(83.3/100) = 7.0$$

c. To compute the exponentially smoothed series, we use the following:

$$E_1 = Y_1$$
$$E_t = wY_t + (1 - w)E_{t-1} \text{ for } t > 1$$

Thus, the exponentially smoothed series is:

$$E_1 = 8.0$$
$$E_2 = .6(6.7) + (1 - .6)(8.0) = 7.22$$
$$E_3 = .6(8.2) + (1 - .6)(7.22) = 7.81$$

The rest of the values of the exponentially smoothed series are found in the same manner and are shown in the following table:

Year	Month	Mortgage Appl.	
1991	Jan.	8.0	8.00
	Feb.	6.7	7.22
	Mar.	8.2	7.81
	Apr.	9.4	8.76
	May	10.5	9.81
	June	8.8	9.20
	July	10.1	9.74
	Aug.	7.4	8.34
	Sep.	7.1	7.59
	Oct.	8.7	8.26
	Nov.	6.6	7.26
	Dec.	6.5	6.81
1992	Jan.	7.2	7.04
	Feb.	7.5	7.32
	Mar.	10.1	8.99
	Apr.	9.4	9.23
	May	7.9	8.43
	June	7.7	7.99
	July	8.8	8.48
	Aug.	7.5	7.89
	Sep.	7.1	7.42
	Oct.	7.3	7.35
	Nov.	7.1	7.20
	Dec.	6.5	6.78
1993	Jan.	5.7	6.13
	Feb.	7.0	6.65
	Mar.	7.7	7.28
	Apr.	8.2	7.83
	May	7.7	7.75
	June	8.4	8.14
	July	8.3	8.24
	Aug.	8.6	8.45
	Sep.	7.4	7.82
	Oct.	8.8	8.41
	Nov.	9.2	8.88
	Dec.	7.1	7.81

The plot of the exponentially smoothed series is:

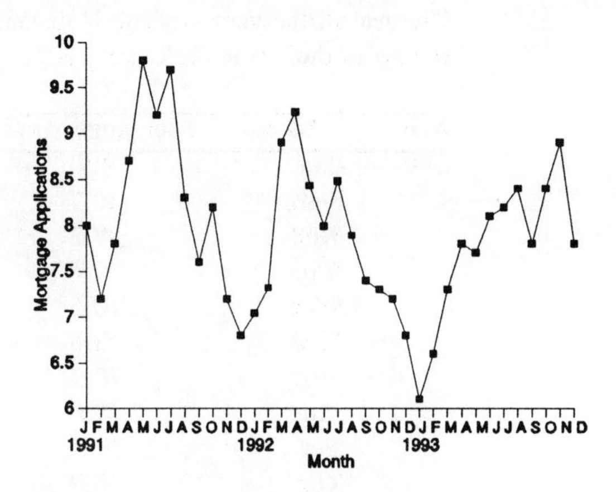

d. The time period January, 1994 is one step from the last observed time series value. To forecast this value, we need to use the following:

$$F_t = wY_n + (1 - w)E_n$$

Thus: $F_{J,1994} = .6(7.1) + (1 - .6)(7.81) = 7.38$

e. To compute the Holt-Winters series with trend and seasonal components, we use the following:

$$E_t = \begin{cases} y_2,\ t = 2 \\ wy_t + (1 - w)(E_{t-1} + T_{t-1}),\ t = 3, 4, \ldots P + 2 \text{ where } P = 12 \\ w(y_t/S_{t-P)} + (1 - w)(E_{t-1} + T_{t-1}),\ t > P + 2 \end{cases}$$

$$T_t = \begin{cases} y_2 - y_1,\ t = 2 \\ v(E_t - E_{t-1}) + (1 - v)T_{t-1},\ t > 2 \end{cases}$$

$$S_t = \begin{cases} y_t/E_t,\ t = 2, 3, \ldots P + 2 \\ u(y_t/E_t) + (1 - u)S_{t-P},\ t > P + 2 \end{cases}$$

Thus, the Holt-Winters series is:

$E_2 = 6.7$
$E_3 = .6(8.2) + (1 - .6)(6.7 + (-1.3) = 7.08$
$E_4 = .6(9.4) + (1 - .6)(7.08 + (-.12) = 8.42$

$T_2 = 6.7 - 8.0 = -1.3$
$T_3 = .7(7.08 - 6.7) + (1 - .7)(-1.3) = -.12$
$T_4 = .7(8.42 - 7.08) + 1 - .7)(-.12) = .09$

$S_2 = y_2/E_2 = 6.7/6.7 = 1.00$
$S_3 = y_3/E_3 = 8.2/7.08 = 1.158$
$S_4 = y_4/E_4 = 7.08/8.42 = .841$

$E_{15} = .6(10.1/1.158) + (1 - .6)(7.19 + .33) = 8.24$
$T_{15} = .7(8.24 - 7.19) + (1 - .7)(.33) = .83$
$S_{15} = .5(y_{15}/E_{15}) + (1 - .5)S_3 = .5(10.1/8.24) + (1 - .5)1.158 = 1.192$

The rest of the values of the Holt-Winters series are found in a similar manner and are shown in the following table:

Year	Month	Mortgage Appl.	E_t	T_t	S_t
1991	Jan.	8.0			
	Feb.	6.7	6.70	−1.30	1.000
	Mar.	8.2	7.08	−0.12	1.158
	Apr.	9.4	8.42	0.90	1.116
	May	10.5	10.03	1.40	1.047
	June	8.8	9.85	0.29	0.893
	July	10.1	10.12	0.27	0.998
	Aug.	7.4	8.60	−0.98	0.861
	Sep.	7.1	7.31	−1.20	0.972
	Oct.	8.7	7.66	−0.11	1.135
	Nov.	6.6	6.98	−0.51	0.945
	Dec.	6.5	6.49	−0.50	1.002
1992	Jan.	7.2	6.72	0.01	1.072
	Feb.	7.5	7.19	0.33	1.043
	Mar.	10.1	8.24	0.84	1.192
	Apr.	9.4	8.69	0.56	1.099
	May	7.9	8.23	−0.15	1.004
	June	7.7	8.40	0.08	0.905
	July	8.8	8.68	0.22	1.006
	Aug.	7.5	8.79	0.14	0.857
	Sep.	7.1	7.95	−0.54	0.932
	Oct.	7.3	6.82	−0.95	1.103
	Nov.	7.1	6.85	−0.26	0.991
	Dec.	6.5	6.53	−0.31	0.999
1993	Jan.	5.7	5.68	−0.69	1.038
	Feb.	7.0	6.02	0.04	1.103
	Mar.	7.7	6.30	0.20	1.207
	Apr.	8.2	7.08	0.61	1.129
	May	7.7	7.68	0.60	1.003
	June	8.4	8.88	1.02	0.925
	July	8.3	8.91	0.33	0.969
	Aug.	8.6	9.72	0.66	0.871
	Sep.	7.4	8.91	−0.36	0.881
	Oct.	8.8	8.21	−0.60	1.087
	Nov.	9.2	8.62	0.10	1.029
	Dec.	7.1	7.75	−0.57	0.957

The forecasts using the Holt-Winters series are:

$$F_t = (E_n + kT_n)S_{n+1-P}, \text{ for } t = n + k; \ k = 1, 2, \ldots ; \ (P = 12)$$

Thus: $F_{J,1994} = (7.75 + (-.57))(1.038) = 7.45$

f. The proposed time series model that accounts for a secular trend, seasonal variation, and residual autocorrelation is:

$$y_t = \beta_0 + \beta_1 t + \beta_2 M_1 + \beta_3 M_2 + \cdots + \beta_{12} M_{11} + \phi R_{t-1} + \epsilon_t \text{ where}$$

$$M_1 = \begin{cases} 1 & \text{if February} \\ 0 & \text{if not} \end{cases} \qquad M_2 = \begin{cases} 1 & \text{if March} \\ 0 & \text{if not} \end{cases}$$

$$\ldots \quad M_{11} = \begin{cases} 1 & \text{if December} \\ 0 & \text{if not} \end{cases}$$

g. The SAS output from fitting the above model is:

```
                         Autoreg Procedure

      Dependent Variable = Y

                  Ordinary Least Squares Estimates

      SSE                 20.21292  DFE                    23
      MSE                 0.878822  Root MSE         0.937455
      SBC                 127.9702  AIC              107.3845
      Reg Rsq               0.5380  Total Rsq          0.5380
      Durbin-Watson         1.3615

Variable     DF       B Value   Std Error   t Ratio    Approx Prob

Intercept     1    7.02500000     0.66288    10.598         0.0001
T             1   -0.01354167     0.01595    -0.849         0.4045
M1            1    0.11770833     0.78527     0.150         0.8822
M2            1    0.23125000     0.78186     0.296         0.7701
M3            1    1.84479167     0.77877     2.369         0.0266
M4            1    2.19166667     0.77599     2.824         0.0096
M5            1    1.90520833     0.77353     2.463         0.0217
M6            1    1.51875000     0.77139     1.969         0.0611
M7            1    2.29895833     0.76957     2.987         0.0066
M8            1    1.07916667     0.76808     1.405         0.1734
M9            1    0.45937500     0.76692     0.599         0.5550
M10           1    1.53958333     0.76609     2.010         0.0563
M11           1    0.91979167     0.76560     1.201         0.2418

                   Estimates of Autocorrelations

                  Lag   Covariance    Correlation

                   0       0.56147       1.000000
                   1      0.164332       0.292681

                        Autoreg Procedure

                   Preliminary MSE = 0.513373

       Estimates of the Autoregressive Parameters

          Lag  Coefficient     Std Error       t Ratio
            1  -0.29268104    0.20386468     -1.435663

                      Yule-Walker Estimates

      SSE                 18.36697  DFE                    22
      MSE                 0.834862  Root MSE         0.913708
      SBC                 128.1957  AIC              106.0264
      Reg Rsq               0.5032  Total Rsq          0.5802
      Durbin-Watson         1.9995

Variable     DF       B Value    Std Error   t Ratio    Approx Prob

Intercept     1    7.12607685      0.72781     9.791         0.0001
T             1   -0.01445376      0.02118    -0.682         0.5021
M1            1    0.07670878      0.71330     0.108         0.9153
M2            1    0.15705585      0.77353     0.203         0.8410
M3            1    1.76152803      0.78638     2.240         0.0355
M4            1    2.10639646      0.78726     2.676         0.0138
M5            1    1.82000540      0.78512     2.318         0.0301
M6            1    1.43424408      0.78240     1.833         0.0804
M7            1    2.21541155      0.77953     2.842         0.0095
M8            1    0.99692153      0.77574     1.285         0.2121
```

```
M9              1   0.37943995      0.76762     0.494          0.6260
M10             1   1.46535650      0.74295     1.972          0.0613
M11             1   0.86286951      0.65394     1.319          0.2006
```

h. For January, 1994, $t = 37$:

$$\hat{R}_{36} = Y_{36} - \hat{Y}_{36} = 7.1 - (7.1261 - .01445(36) + .8629(1)) = -.3688$$

$$\begin{aligned}\hat{R}_{37} &= 7.1261 - .01445(37) + .29(\hat{R}_{36}) \\ &= 7.1261 - .01445(37) + .29(-.3688) = 6.483\end{aligned}$$

The form of the prediction interval is:

$$\hat{Y}_{n+m} \pm 2\sqrt{\text{MSE}\left(1 + \hat{\phi}^2 + \hat{\phi}^4 + \cdots + \hat{\phi}^{2(m-1)}\right)}$$

where m is the number of steps ahead.

For $t = 37$, $m = 1$. The approximate 95% prediction interval is:

$$\hat{Y}_{37} \pm 2\sqrt{\text{MSE}} \Rightarrow 6.483 \pm 2(.9137) \Rightarrow 6.483 \pm 1.827 \Rightarrow (4.656, 8.310)$$

9.39 a. Suppose we choose the exponentially smoothed series with $w = .6$. The exponentially smoothed value for the first quarter of 1978 is:

$$E_{1978,\text{I}} = y_{1978,\text{I}} = 1{,}711$$

For each subsequent quarter, we apply the formula:

$$E_t = wy_t + (1 - w)E_{t-1}$$

where $w = .6$ is the smoothing constant.

$$E_{1978,\text{II}} = .6y_{1978,\text{II}} + .4E_{1978,\text{I}} = .6(4065) + .4(1711) = 3123.40$$
$$E_{1978,\text{III}} = .6y_{1978,\text{III}} + .4E_{1978,\text{II}} = .6(5787) + .4(3123.40) = 4721.56$$
$$\vdots$$

The complete exponentially smoothed series is given in the following table:

Year	Quarter	y_t	$E_t(w = .6)$
1978	I	1,711	1,711.00
	II	4,065	3,123.40
	III	5,787	4,721.56
	IV	5,019	4,900.02
1979	I	5,459	5,235.41
	II	9,184	7,604.56
	III	12,168	10,342.63
	IV	11,842	11,242.25
1980	I	13,730	12,734.90
	II	14,964	14,072.36
	III	18,058	16,463.74
	IV	21,393	19,421.30

The forecasted value for the first quarter of 1981 is:

$$F_t = wy_n + (1 - w)E_n$$

where y_n and E_n are the observed value and exponentially smoothed value, respectively, for the last time period in the series. Therefore,

$$F_{1981,I} = .6y_{1980,IV} + .4E_{1980,IV} = .6(21{,}393) + .4(19{,}421.30) = 20{,}604.32$$

b. The model is $y_t = \beta_0 + \beta_1 t + \beta_2 Q_1 + \beta_3 Q_2 + \beta_4 Q_3 + \phi R_{t-1} + \epsilon_t$

where $Q_1 = \begin{cases} 1 & \text{if quarter I} \\ 0 & \text{otherwise} \end{cases}$ $Q_2 = \begin{cases} 1 & \text{if quarter II} \\ 0 & \text{otherwise} \end{cases}$ $Q_3 = \begin{cases} 1 & \text{if quarter III} \\ 0 & \text{otherwise} \end{cases}$

Using SAS, the printout is as follows:

```
                A U T O R E G   P R O C E D U R E

DEPENDENT VARIABLE = Y

                ORDINARY LEAST SQUARES ESTIMATES

        SSE             13384988    DFE                 7
        MSE              1912141    ROOT MSE     1382.802
        SBC             213.5759    AIC          211.1514
        REG RSQ           0.9672    TOTAL RSQ      0.9672
        DURBIN-WATSON     1.1935

VARIABLE DF          B VALUE       STD ERROR   T RATIO APPROX PROB

INTERCPT  1       -139.41667      1262.31965    -0.110      0.9152
T         1       1611.34375       122.22357    13.184      0.0001
Q1        1       -950.63542      1187.10073    -0.801      0.4496
Q2        1       -124.31250      1155.21216    -0.108      0.9173
Q3        1        864.34375      1135.64929     0.761      0.4715

                  ESTIMATES OF AUTOCORRELATIONS

                   LAG  COVARIANCE  CORRELATION
                    0      1115416     1.000000
                    1       189842     0.170198

                  PRELIMINARY MSE=       1083105

              ESTIMATES OF THE AUTOREGRESSIVE PARAMETERS
        LAG     COEFFICIENT       STD ERROR         T RATIO
          1     -0.17019816      0.40229189       -0.423071

                       YULE-WALKER ESTIMATES

        SSE             12751903    DFE                 6
        MSE              2125317    ROOT MSE     1457.847
        SBC             215.5088    AIC          212.5993
        REG RSQ           0.9600    TOTAL RSQ      0.9688

VARIABLE DF          B VALUE       STD ERROR   T RATIO APPROX PROB

INTERCPT  1       -173.93270      1426.70667    -0.122      0.9069
T         1       1623.68391       147.47471    11.010      0.0001
Q1        1       -859.24861      1190.07150    -0.722      0.4974
Q2        1       -141.86720      1204.34948    -0.118      0.9101
Q3        1        826.73264      1103.61385     0.749      0.4821
```

The fitted model is:

$$\hat{y}_t = -173.933 + 1623.684t - 859.249Q_1 - 141.867Q_2 + 826.733Q_3 + .1702\hat{R}_{t-1}$$

c. To forecast the employment in quarter I, 1981, $t = 13$

First, we calculate the estimated residual for the last observation; i.e., $\hat{R}_{12}$:

$$\hat{R}_{12} = y_{12} - \hat{y}_{12} = y_{12} - [\hat{\beta}_0 + \hat{\beta}_1(12)]$$
$$= 21{,}393 - [-173.933 + 1623.684(12)] = 2082.725$$

For $t = 13$, $\hat{R}_{13} = .1702\hat{R}_{12} = .1702(2082.725) = 354.480$

$$\hat{y}_{1981,\text{I}} = -173.933 + 1623.684(13) - 859.249(1) + 354.480 = 20{,}429.19$$

An approximate 95% prediction interval for $\hat{y}_{13}$ is calculated as follows, where MSE = 2,125,317 is obtained from the SAS output.

$$t = 13 \text{ (1981,I)}: \quad \hat{y}_{13} \pm 2\sqrt{\text{MSE}} \Rightarrow 20{,}429.19 \pm 2\sqrt{2{,}125{,}317}$$
$$\Rightarrow 20{,}429.19 \pm 2915.69$$
$$\Rightarrow (17{,}513.5, 23{,}344.88)$$

d. The actual value for quarter I of 1981, 22,772, does fall in the interval.

e. It would be rather risky to use this model to forecast quarterly employment in 1995 because it is so far away from the last observed value.

9.41 a. Using SAS, the output for fitting the data is:

```
Model:  MODEL 1
Dependent Variable Y

                        Analysis of Variance

                          Sum of          Mean
Source          DF       Squares        Square   F Value     Prob>F

Model            1    4737.97221    4737.79221    52.519     0.0001
Error           19    1090.87446      57.41445
C Total         20    5828.66667

     Root MSE          7.57723   R-square     0.8128
     Dep Mean        187.33333   Adj R-sq     0.8030
     C.V.              4.04479

                        Parameter Estimates

                      Parameter      Standard      T for H0:
 Variable      DF      Estimate         Error    Parameter = 0     Prob > |T|

 INTERCEP       1    160.047619    3.42874484           46.678         0.0001
 T              1      2.480519    0.27306449            9.084         0.0001

Durbin-Watson D                0.225
(For Number of Obs.)              21
1st Order Autocorrelation      0.792
```

Obs	Dep Var Y	Predict Value	Residual
1	149.0	162.5	-13.5281
2	156.0	165.0	-9.0087
3	161.0	167.5	-6.4892
4	164.0	170.0	-5.9697
5	171.0	172.5	-1.4502
6	179.0	174.9	4.0693
7	184.0	177.4	6.5887
8	194.0	179.9	14.1082
9	194.0	182.4	11.6277
10	196.0	184.9	11.1472
11	196.0	187.3	8.6667
12	193.0	189.8	3.1861
13	194.0	192.3	17056
14	197.0	194.8	2.2251
15	195.0	197.3	-2.2554
16	197.0	199.7	-2.7359
17	199.0	202.2	-3.2165
18	204.0	204.7	-0.6970
19	202.0	207.2	-5.1775
20	202.0	209.7	-7.6580
21	207.0	212.1	-5.1385

Sum of Residuals	0
Sum of Squared Residuals	1090.8745
Predicted Resid SS (Press)	1356.6129

The fitted regression line is $\hat{y} = 160.048 + 2.481t$.

b. The plot of the least squares line is:

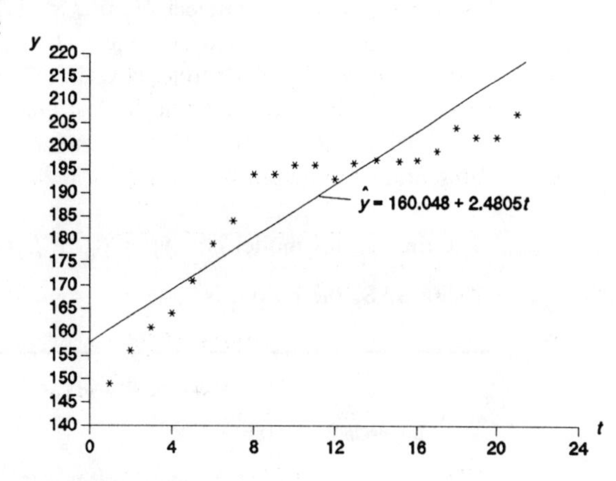

Using the fitted regression line, the forecast for y_{22} is:

$$\hat{y} = 160.048 + 2.481(22) = 214.63$$

Since this forecast is for a value outside the observed range, it is not particularly reliable.

c. The residuals are shown in the output in part (a). The plot of the residuals is:

There appears to be evidence of autocorrelation. There is a U-shape to the residuals.

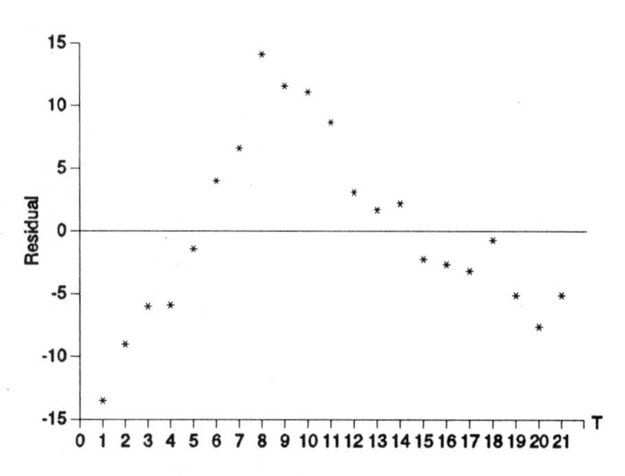

d. We could test for the presence of autocorrelation using the Durbin Watson statistic. From the printout in part (a), the value of the Durbin Watson statistic is .225.

H_0: No autocorrelation
H_a: Positive or negative autocorrelation

Test statistic: $d = .225$

Rejection region: $\alpha = .10$, $k = 1$, $n = 21$, $d_{L,.05} = 1.22$; $d_{U,.05} = 1.41$
Reject H_0 if $d < 1.22$ or $(4 - d) < 1.22$
Do not reject H_0 if $d > 1.41$ or $(4 - d) > 1.41$
Inconclusive if $1.22 < d < 1.41$ or $1.22 < (4 - d) < 1.41$

Conclusion: Reject H_0. There is sufficient evidence to indicate that autocorrelation is present at $\alpha = .10$.

e. The time series model is: $y_t = \beta_0 + \beta_1 t + \phi R_{t-1} + \epsilon_t$

Using SAS, the output is:

```
                    Autoreg Procedure

     Dependent Variable = Y

                Ordinary Least Squares Estimates

     SSE                1090.874  DFE                19
     MSE                57.41445  Root MSE     7.577232
     SBC                148.6389  AIC          146.5499
     Reg Rsq              0.8128  Total Rsq      0.8128
     Durbin-Watson        0.2249

Variable     DF      B Value   Std Error   t Ratio   Approx Prob

Intercept     1   160.047619      3.4287    46.678        0.0001
T             1     2.480519      0.2731     9.084        0.0001

                Estimates of Autocorrelations

               Lag   Covariance   Correlation

                0       51.9464      1.000000
                1       41.1194      0.791574

               Preliminary MSE = 19.39738
```

```
Estimates of the Autoregressive Parameters

   Lag  Coefficient    Std Error     t Ratio
     1  -0.79157350  0.14403151   -5.495835

                     Autoreg Procedure

                   Yule-Walker Estimates

SSE                    263.3938  DFE                  18
MSE                    14.63299  Root MSE       3.825309
SBC                    122.8257  AIC            119.6922
Reg Rsq                  0.6994  Total Rsq        0.9548
Durbin-Watson            1.2482
```

Variable	DF	B Value	Std Error	t Ratio	Approx Prob
Intercept	1	154.830097	5.7705	26.831	0.0001
T	1	2.729462	0.4217	6.472	0.0001

The fitted time series model is:

$$\hat{y}_t = 154.83 + 2.729t + .792\hat{R}_{t-1}$$

For y_{22} $t = 22$:

$$\hat{R}_{21} = Y_{21} - \hat{Y}_{21} = 207 - (154.83 + 2.729(21)) = -5.139$$

$$\begin{aligned}\hat{Y}_{22} &= 154.83 + 2.729(22) + .792\hat{R}_{21} \\ &= 154.83 + 2.729(22) + .792(-5.139) = 210.80\end{aligned}$$

The form of the prediction interval is:

$$\hat{Y}_{n+m} \pm 2\sqrt{\text{MSE}\left(1 + \hat{\phi}^2 + \hat{\phi}^4 + \cdots + \hat{\phi}^{2(m-1)}\right)}$$

where m is the number of steps ahead.

For $t = 22$, $m = 1$. The approximate 95% prediction interval is:

$$\hat{Y}_{22} \pm 2\sqrt{\text{MSE}} \Rightarrow 210.80 \pm 2(3.825) \Rightarrow 210.80 \pm 7.650 \Rightarrow (203.15, 218.45)$$

This forecast is preferred over that in part (b) because it takes into account that observations are correlated with each other.

9.43 a.

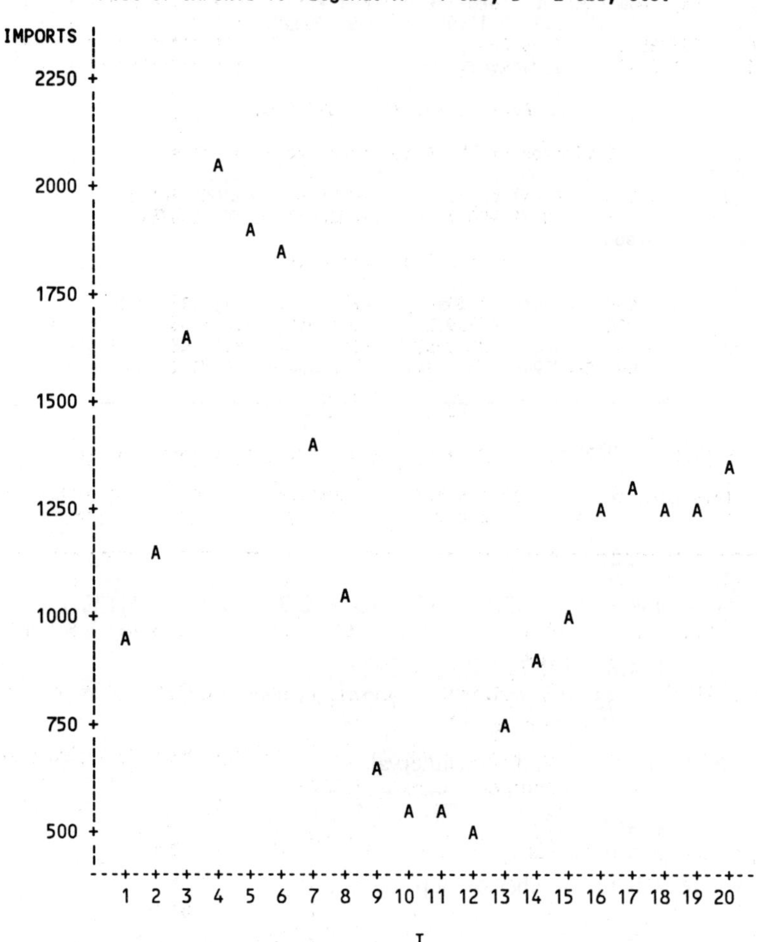

b. $Y_t = \beta_0 + \beta_1 t + \phi R_{t-1} + \epsilon_t$

c. Using SAS to fit the model, we get the following output:

```
                         Autoreg Procedure

Dependent Variable = IMPORTS

                   Ordinary Least Squares Estimates

           SSE             3742583    DFE               18
           MSE            207921.3    Root MSE    455.9838
           SBC            305.5401    AIC         303.5486
           Reg Rsq          0.0743    Total Rsq     0.0743
           Durbin-Watson    0.3265

     Variable     DF        B Value    Std Error    t Ratio Approx Prob

     Intercept     1     1384.59474       211.82      6.537      0.0001
     T             1      -21.25188        17.68     -1.202      0.2450
```

```
                     Estimates of Autocorrelations

Lag  Covariance  Correlation -1 9 8 7 6 5 4 3 2 1 0 1 2 3 4 5 6 7 8 9 1

  0    187129.1     1.000000 |                    |********************|
  1      148197     0.791950 |                    |****************    |

                      Preliminary MSE = 69764.48

               Estimates of the Autoregressive Parameters

               Lag    Coefficient      Std Error         t Ratio
                 1    -0.79195037     0.14808873       -5.347810

                         Yule-Walker Estimates

               SSE              1134567    DFE               17
               MSE             66739.21    Root MSE    258.3393
               SBC              285.652    AIC         282.6648
               Reg Rsq           0.0016    Total Rsq     0.7194

                          Autoreg Procedure

     Variable     DF          B Value   Std Error   t Ratio Approx Prob

     Intercept     1       1101.83127      394.78     2.791      0.0125
     T             1          4.91781       30.12     0.163      0.8722
```

$\hat{\beta}_0 = 1{,}101.83$: This is the estimate of where the regression line will cross the y_t axis.

$\hat{\beta}_1 = 4.918$: The estimated increase in annual OPEC oil imports is 4.918 every time period.

$\hat{\phi} = .7920$: Since $\hat{\phi}$ is positive, it implies that the time series residuals are positively correlated.

d. The prediction equation is $\hat{y}_t = 1{,}101.83 + 4.918t + 0.7920\hat{R}_{t-1}$.

e. For 1993: $\hat{R}_{20} = y_{20} - \hat{y}_{20} = 1{,}339 - (1{,}101.83 + 4.918t)$
$= 1{,}339 - (1{,}101.83 + 4.918(20)) = 138.81$

Thus, $\hat{y}_{21} = 1{,}101.83 + 4.918(21) + 0.7920\hat{R}_{20}$
$= 1{,}205.11 + 0.7920(138.81) = 1{,}315.04$

The approximate 95% prediction interval is:

$$\hat{y}_{21} \pm 2\sqrt{\text{MSE}} \Rightarrow 1{,}315.04 \pm 2\sqrt{66{,}739.21} \Rightarrow 1{,}315.04 \pm 516.68 \Rightarrow (798.36, 1{,}831.72)$$

10 Principles of Experimental Design

10.1 The two factors that affect the quantity of information in an experiment are noise (variability) and volume (n).

10.3 a. This experiment involves a single factor, work scheduling, at three levels, flextime, staggered starting hours, and fixed hours. Since work scheduling is the only factor, these levels represent the treatments.

b. To assign treatments in a completely random manner, with equal numbers of workers in each treatment, number the 60 workers from 1 to 60. Use Table 7 in Appendix C to select two-digit numbers, discarding those that are larger than 60 or are identical, until there is a total of 40 two-digit numbers. The workers who have been assigned the first 20 numbers in the sequence are assigned to flextime, the second group of 20 workers are assigned to staggered starting times, and the remaining workers are assigned to fixed hours.

c. The linear model is $E(y) = \beta_0 + \beta_1 x_1 + \beta_2 x_2$

where $x_1 = \begin{cases} 1 & \text{if flextime} \\ 0 & \text{otherwise} \end{cases}$ and $x_2 = \begin{cases} 1 & \text{if staggered} \\ 0 & \text{otherwise} \end{cases}$

10.5 a. The model for the first observation for Appraiser B ($x_1 = 0, x_2 = 1, x_3 = 0$) is:

$$y_{B1} = \beta_0 + \beta_2 + \beta_4 + \epsilon_{B1}$$

The models for the rest of the observations for Appraiser B are:

$$\begin{aligned}
y_{B2} &= \beta_0 + \beta_2 + \beta_5 + \epsilon_{B2} \\
y_{B3} &= \beta_0 + \beta_2 + \beta_6 + \epsilon_{B3} \\
y_{B4} &= \beta_0 + \beta_2 + \beta_7 + \epsilon_{B4} \\
y_{B5} &= \beta_0 + \beta_2 + \beta_8 + \epsilon_{B5} \\
y_{B6} &= \beta_0 + \beta_2 + \beta_9 + \epsilon_{B6} \\
y_{B7} &= \beta_0 + \beta_2 + \beta_{10} + \epsilon_{B7} \\
y_{B8} &= \beta_0 + \beta_2 + \beta_{11} + \epsilon_{B8} \\
y_{B9} &= \beta_0 + \beta_2 + \beta_{12} + \epsilon_{B9} \\
y_{B10} &= \beta_0 + \beta_2 + \epsilon_{B10}
\end{aligned}$$

The average of the 10 observations for Appraiser B is:

$$\bar{y}_B = \beta_0 + \beta_2 + \frac{\beta_4 + \beta_5 + \cdots + \beta_{12}}{10} + \bar{\epsilon}_B$$

b. The models for the observations for Appraiser D ($x_1 = 0, x_2 = 1, x_3 = 0$) are:

$$\begin{aligned}
y_{D1} &= \beta_0 + \beta_4 + \epsilon_{D1} \\
y_{D2} &= \beta_0 + \beta_5 + \epsilon_{D2} \\
y_{D3} &= \beta_0 + \beta_6 + \epsilon_{D3} \\
y_{D4} &= \beta_0 + \beta_7 + \epsilon_{D4} \\
y_{D5} &= \beta_0 + \beta_8 + \epsilon_{D5} \\
y_{D6} &= \beta_0 + \beta_9 + \epsilon_{D6} \\
y_{D7} &= \beta_0 + \beta_{10} + \epsilon_{D7} \\
y_{D8} &= \beta_0 + \beta_{11} + \epsilon_{D8} \\
y_{D9} &= \beta_0 + \beta_{12} + \epsilon_{D9} \\
y_{D10} &= \beta_0 + \epsilon_{D10}
\end{aligned}$$

The average of the 10 observations for Appraiser D is:

$$\bar{y}_D = \beta_0 + \frac{\beta_4 + \beta_5 + \cdots + \beta_{12}}{10} + \bar{\epsilon}_D$$

c. $$(\bar{y}_B - \bar{y}_D) = \left[\beta_0 + \beta_2 + \frac{\beta_4 + \cdots + \beta_{12}}{10} + \bar{\epsilon}_B\right] - \left[\beta_0 + \frac{\beta_4 + \cdots + \beta_{12}}{10} + \bar{\epsilon}_D\right]$$

$$= \beta_2 + (\bar{\epsilon}_B - \bar{\epsilon}_D)$$

10.7 a. Since y will be observed for all factor-level combinations, this is a complete 3×3 factorial experiment.

b. The factors are pay rate (quantitative) and length of workday (quantitative).

c. The treatments will include all $3 \times 3 = 9$ factor-level combinations:

$$P_1L_1, P_1L_2, P_1L_3, P_2L_1, P_2L_2, P_2L_3, P_3L_1, P_3L_2, P_3L_3.$$

10.9 a. This is not a complete factorial experiment since the treatments do not include all $3 \times 3 = 9$ factor-level combinations.

b. Interaction between two factors implies the effect of one factor on the dependent variable depends on the level of the second factor. There are 3 levels of factor A at B_1. However, there is only 1 level of A at B_2 and only 1 level of A at B_3. Thus, we cannot measure the effect of A at levels B_2 and B_3. Therefore, we cannot determine if the effect of A differs at different levels of B.

10.11 Let x_1 be the quantitative factor A,

$$x_2 = \begin{cases} 1 & \text{if Factor } B \text{ at level 1} \\ 0 & \text{otherwise} \end{cases} \qquad x_3 = \begin{cases} 1 & \text{if Factor } B \text{ at level 2} \\ 0 & \text{otherwise} \end{cases}$$

$$x_4 = \begin{cases} 1 & \text{if Factor } C \text{ at level 1} \\ 0 & \text{otherwise} \end{cases} \qquad x_5 = \begin{cases} 1 & \text{if Factor } C \text{ at level 2} \\ 0 & \text{otherwise} \end{cases}$$

The model is:

$$E(y) = \beta_0 + \beta_1x_1 + \beta_2x_2 + \beta_3x_3 + \beta_4x_4 + \beta_5x_5 + \beta_6x_1x_2 + \beta_7x_1x_3 + \beta_8x_1x_4 + \beta_9x_1x_5 + \beta_{10}x_2x_4 + \beta_{11}x_2x_5 + \beta_{12}x_3x_4 + \beta_{13}x_3x_5 + \beta_{14}x_1x_2x_4 + \beta_{15}x_1x_2x_5 + \beta_{16}x_1x_3x_4 + \beta_{17}x_1x_3x_5$$

10.13 If a randomized block design is used to investigate the effect of two qualitative factors, there will be only one observation per treatment. There will not be enough degrees of freedom (or observations) to test for interaction between the two factors.

10.15 To solve for the number of replicates, r, we want to solve the equation:

$$t_{\alpha/2}\frac{s_{\hat{\beta}_3}}{\sqrt{r}} = B$$

We know the estimate of $s_{\hat{\beta}_3}$ is 3 and $B = 2$. For $\alpha = .10$, $\alpha/2 = .05$.

To determine the value of $t_{.05}$, we need to know the degrees of freedom. We know the degrees of freedom for t will be df (Error) $= n - 4 = 4r - 4 = 4(r - 1)$. At minimum, we require 2 replicates, so we will start with $r = 2$. Thus, with df $= 4$, $t_{.05} = 2.132$. Substituting into the formula, we get:

$$2.132\frac{3}{\sqrt{r^2}} = 2 \Rightarrow r = \frac{2.132^2(3^2)}{2^2} = 10.23$$

Since this number is quite a bit larger than the 2 replicates that we used to get the original t value, we will redo the problem using the $t_{.05}$ with df $= 4(r - 1) = 4(10 - 1) = 36$. This value is $t_{.05} \approx 1.69$.

$$1.690\frac{3}{\sqrt{r^2}} = 2 \Rightarrow r = \frac{1.690^2(3^2)}{2^2} = 6.43 \approx 7$$

Using 7 as the number of replicates, df $= 4(r - 1) = 24$ and $t_{.05} = 1.711$.

$$1.711\frac{3}{\sqrt{r^2}} = 2 \Rightarrow r = \frac{1.711^2(3^2)}{2^2} = 6.59 \approx 7$$

Thus, we should use either 6 or 7 replicates. To be conservative, we should use 7 replicates.

10.17 The quantity of information in an experiment may be controlled by following these four steps:

Step 1 Select the factors to be included and define the parameters of interest in the experiment.
Step 2 Choose the factor-level combinations (treatments) to be included in the experiment.
Step 3 Determine the number of observations per treatment.
Step 4 Decide how to allocate experimental units to the treatments.

10.19 The variation produced by extraneous variables can be reduced in step 4 by the process known as blocking. We observe all the treatments within relatively homogeneous blocks of experimental material. By assigning the treatments within blocks, we may decrease the noise (variability) in an experiment.

10.21 Factor 1 contains 2 levels – A_1 and A_2.
Factor 2 contains 4 levels – B_1, B_2, B_3 and B_4.

There are $2 \times 4 = 8$ possible factor-level combinations. They are:

$$A_1B_1, A_1B_2, A_1B_3, A_1B_4, A_2B_1, A_2B_2, A_2B_3, A_2B_4$$

10.23 Let x_1 and x_2 correspond to the 2 quantitative factors at 2 levels.

$$x_3 = \begin{cases} 1 & \text{if Factor } C \text{ at level 1} \\ 0 & \text{otherwise} \end{cases} \qquad x_4 = \begin{cases} 1 & \text{if Factor } C \text{ at level 2} \\ 0 & \text{otherwise} \end{cases}$$

$$x_5 = \begin{cases} 1 & \text{if Factor } C \text{ at level 3} \\ 0 & \text{otherwise} \end{cases}$$

The model with all main effects but no interactions is:

$$E(y) = \beta_0 + \beta_1x_1 + \beta_2x_2 + \beta_3x_3 + \beta_4x_4 + \beta_5x_5$$

The number of degrees of freedom for estimating σ^2 is $n - (k + 1) = 16 - (5 + 1) = 10$.

10.25 a. The two factors are Sex and Weight.

b. Sex is a qualitative factor.
Weight is a quantitative factor.

c. The four treatments are:

(M, L); (M, H); (F, L); (F, H)

11 The Analysis of Variance for Designed Experiments

11.1 Recall in Exercise 10.3 we first number the workers from 1 to 60. Then using Table 7, Appendix C, we will select 2-digit numbers until we have 40 different 2-digit numbers between 1 and 60. Those workers assigned the first 20 numbers will be assigned to flextime, those assigned to the next 20 numbers will be assigned to staggered starting times, and those remaining will be assigned to fixed hours.

Starting in column 6, row 18, using the last two digits, and going down the column, we get the following random numbers:

46, 13, 6, 5, 56, 20, 51, 49, 27, 48, 40, 26, 54, 8, 17, 31, 38, 34, 58, 32, 16, 57, 42, 60, 33, 21, 23, 47, 19, 15, 2, 28, 41, 37, 9, 53, 22, 29, 52, 12

Thus, those assigned to flextime are workers numbered 5, 6, 8, 13, 17, 20, 26, 27, 31, 32, 34, 38, 40, 46, 48, 49, 51, 54, 56, and 58.

Those assigned to staggered starting times are workers numbered 2, 9, 12, 15, 16, 19, 21, 22, 23, 28, 29, 33, 37, 41, 42, 47, 52, 53, 57, and 60.

Those assigned to fixed hours are workers numbered 1, 3, 4, 7, 10, 11, 14, 18, 24, 25, 30, 35, 36, 39, 43, 44, 45, 50, 55, and 59.

11.3 a. The linear model is: $E(y) = \beta_0 + \beta_1 x$ where $x = \begin{cases} 1 & \text{if Treatment 1} \\ 0 & \text{otherwise} \end{cases}$

b. Some preliminary calculations are:

$$n = 13 \quad \sum x = 7 \quad \sum y = 128 \quad \sum xy = 64$$
$$\sum x^2 = 7 \quad \sum y^2 = 1294$$

$$SS_{xx} = 7 - \frac{7^2}{13} = 3.230769231 \qquad SS_{xy} = 64 - \frac{7(128)}{13} = -4.92307692$$

$$SS_{yy} = 1294 - \frac{128^2}{13} = 33.692308$$

$$\hat{\beta}_1 = \frac{SS_{xy}}{SS_{xx}} = \frac{-4.92307692}{3.230769231} = -1.523809523 \approx -1.524$$

$$\hat{\beta}_0 = \bar{y} - \hat{\beta}_1\bar{x} = \frac{128}{13} - (-1.523809523)\left(\frac{7}{13}\right) = 10.666666667 \approx 10.667$$

The fitted model is $\hat{y} = 10.667 - 1.524x$.

$$SSE = SS_{yy} - \hat{\beta}_1 SS_{xy} = 33.692308 - (-1.523809523)(-4.92307692) = 26.19047651$$

$$s^2 = \frac{\text{SSE}}{n - 2} = \frac{26.19047651}{13 - 2} = 2.38095241 \qquad s = \sqrt{2.38095241} = 1.543$$

H_0: $\beta_1 = 0$
H_a: $\beta_1 \neq 0$

Test statistic: $t = \dfrac{\hat{\beta}_1}{s_{\hat{\beta}_1}} = \dfrac{-1.524}{\dfrac{1.543}{\sqrt{3.23077}}} = -1.775$

Rejection region: For $\alpha = .05$, df $= n - 2 = 11$, $t_{.025} = 2.201$
Reject H_0 if $t < -2.201$ or $t > 2.201$

Conclusion: Do not reject H_0 at $\alpha = .05$. There is insufficient evidence to indicate a difference in the treatment means.

Note: We could also test the hypothesis using the F test. To calculate the F statistic, we must first calculate r^2.

$$r^2 = 1 - \frac{\text{SSE}}{\text{SS}_{yy}} = 1 - \frac{26.1905}{33.6923} = .2227$$

Test statistic: $F = \dfrac{r^2/k}{(1 - r^2)/[n - (k + 1)]} = \dfrac{.2227/1}{(1 - .2227)/[13 - (1 + 1)]} = 3.15$

Rejection region: For $\alpha = .05$, $\nu_1 = k = 1$, $\nu_2 = n - (k + 1) = 11$
$F_{.05} = 4.84$. Reject H_0 if $F > 4.84$

The conclusion is the same as using the t test.

11.5 a. Some preliminary calculations:

$$\bar{y}_1 = \frac{64}{7} = 9.143 \qquad \bar{y}_2 = \frac{64}{6} = 10.667$$

$$s_1^2 = \frac{596 - \frac{64^2}{7}}{7 - 1} = 1.8095 \qquad s_2^2 = \frac{698 - \frac{64^2}{6}}{6 - 1} = 3.0667$$

$$s_p^2 = \frac{(n_1 - 1)s_1^2 + (n_2 - 1)s_2^2}{n_1 + n_2 - 2} = \frac{(7 - 1)1.8095 + (6 - 1)3.0667}{7 + 6 - 2} = 2.3810$$

H_0: $\mu_1 = \mu_2$
H_a: $\mu_1 \neq \mu_2$

Test statistic: $t = \dfrac{\bar{y}_1 - \bar{y}_2}{\sqrt{s_p^2\left(\dfrac{1}{n_1} + \dfrac{1}{n_2}\right)}} = \dfrac{9.143 - 10.667}{\sqrt{2.381\left(\dfrac{1}{7} + \dfrac{1}{6}\right)}} = -1.775$

Rejection region: For $\alpha = .05$, $t_{.025} = 2.201$ where t is based on $(n_1 + n_2 - 2)$ $= 11$ degrees of freedom.
Reject H_0 if $t < -2.201$ or $t > 2.201$

Conclusion: Do not reject H_0 at $\alpha = .05$. There is insufficient evidence to indicate a difference in the treatment means.

b. $t^2 = (-1.775)^2 = 3.15 = F$

c. The analysis of variance F test for comparing two population means is always a two-tailed test with the alternative hypothesis H_a: $\mu_1 \neq \mu_2$

11.7 a. There are 3 treatments in this experiment. They are eating disorder patients, depressed patients, and normal eaters.

b. The null hypothesis is H_0: $\mu_1 = \mu_2 = \mu_3$

c. df(Total) $= n - 1 = 83 - 1 = 82$
df(Error) $= n - k = 83 - 3 = 80$
df(Treatments) $= k - 1 = 3 - 1 = 2$

d. The rejection region requires $\alpha = .01$ in the upper tail of the F distribution with $\nu_1 = k - 1 = 3 - 1 = 2$ and $\nu_2 = n - k = 83 - 3 = 80$. From Table 6, Appendix C, $F_{.01} \approx 4.98$. The rejection region is $F > 4.98$.

Since the observed value of the test statistic falls in the rejection region ($F = 11.5 > 4.98$), H_0 is rejected. There is sufficient evidence to indicate a difference in the mean self-esteem scores among the 3 groups of women at $\alpha = .01$.

11.9 a. This is a completely randomized design. The 45 auditors were randomly assigned to one of the three treatments.

b. The three treatments in this experiment are the three groups: A/R, A/P, and Control.

c. df(Groups) $= k - 1 = 3 - 1 = 2$

df(Error) $= n - k = 45 - 3 = 42$

SSE $=$ SSTotal $-$ SST $= 392.98 - 71.51 = 321.47$

$$\text{MST} = \frac{\text{SST}}{k-1} = \frac{71.51}{3-1} = 35.755$$

$$F = \frac{\text{MST}}{\text{MSE}} = \frac{35.755}{7.65} = 4.67$$

The completed table is:

Source	df	SS	MS	F	p-value
Groups	2	71.51	35.755	4.67	.01
Error	42	321.47	7.65		
Total	44	392.98			

d. To determine if the group means differ, we test:

H_0: $\mu_1 = \mu_2 = \mu_3$
H_a: At least two means differ

The test statistic is $F = 4.67$.

The p-value for this test is .01. Since this value is very small, we would reject H_0. There is sufficient evidence to indicate a difference in the mean number of hours allocated to the accounts receivable among the three groups at $\alpha = .05$.

11.11 H_0: $\mu_1 = \mu_2 = \mu_3 = \mu_4$
H_a: At least two of the four means differ

Test statistic: $F = .10$ (from printout)

Rejection region: Reject H_0 if p-value $< \alpha = .05$
$p = .9566$

Conclusion: Do not reject H_0 at $\alpha = .05$. There is insufficient evidence to indicate a difference in the mean number of trials required for the four groups.

11.13 a. The treatments for this experiment are the five filtering methods: (1) keyword match-word profile, (2) latent semantic indexing (LSI)-word profile, (3) keyword match-document profile, (4) LSI-document profile, and (5) random selection method.

b. The response variable is the relevance rating of the technical memos.

c. H_0: $\mu_1 = \mu_2 = \mu_3 = \mu_4 = \mu_5$
H_a: At least two means differ

Test statistic: $F = 117.5$

Rejection region: $\alpha = .05$, $\nu_1 = 4$, $\nu_2 = 132$, $F_{.05} \approx 2.45$
Reject H_0 if $F > 2.45$.

Conclusion: Reject H_0. There is sufficient evidence (at $\alpha = .05$) to indicate the mean relevance ratings differ among the five filtering methods.

11.15 a. The linear model is:

$$E(y) = \beta_0 + \beta_1 x_1 + \beta_2 x_2$$

where $x_1 = \begin{cases} 1 & \text{if entrepreneurs} \\ 0 & \text{otherwise} \end{cases}$ $\quad x_2 = \begin{cases} 1 & \text{if transferred managers} \\ 0 & \text{otherwise} \end{cases}$

b. For Promoted managers, $x_1 = 0, x_2 = 0$

$$E(y) = \beta_0 = \mu_3 \Rightarrow \hat{\beta}_0 = \bar{y}_3 = 66.97$$

For Entrepreneurs, $x_1 = 1, x_2 = 0$

$$E(y) = \beta_0 + \beta_1 \Rightarrow \mu_1 = \beta_0 + \beta_1 \Rightarrow \beta_1 = \mu_1 - \mu_3$$
$$\Rightarrow \hat{\beta}_1 = \bar{y}_1 - \bar{y}_3 = 71.00 - 66.97 = 4.03$$

For Transferred managers, $x_1 = 0, x_2 = 1$

$$E(y) = \beta_0 + \beta_2 \Rightarrow \mu_2 = \beta_0 + \beta_2 \Rightarrow \beta_2 = \mu_2 - \mu_3$$
$$\Rightarrow \hat{\beta}_2 = \bar{y}_2 - \bar{y}_3 = 72.52 - 66.97 = 5.55$$

The least squares prediction equation is:

$$\hat{y} = 66.97 + 4.03x_1 + 5.55x_2$$

11.17 a. H_0: $\mu_1 = \mu_2$
H_a: $\mu_1 \neq \mu_2$

Test statistic: $F = 2.32$

Rejection region: Reject H_0 if p-value $< \alpha = .10$
$p = .13$

Conclusion: Do not reject H_0 at $\alpha = .10$. There is insufficient evidence to indicate a difference between the mean trustworthiness scores of the two groups of MBA students.

b. H_0: $\mu_1 = \mu_2$
H_a: $\mu_1 < \mu_2$

Test statistic: We know that $t^2 = F$. Thus, the test statistic is $t = \sqrt{2.32} = 1.52$. Since $\bar{y}_1 < \bar{y}_2$, $t = -1.52$.

Rejection region: Reject H_0 if p-value $< \alpha = .10$
Since this is a one-tailed test, $p = .13/2 = .065$

Conclusion: Reject H_0 at $\alpha = .10$. There is sufficient evidence to indicate the mean trustworthiness score of the group of MBA students viewing the Indian sales representative is less than the mean trustworthiness score of the group of MBA students viewing the American sales representative.

11.19 a. The three positions represent the treatments in the experiment.

b. The ten college students represent the blocks in the experiment.

c. The medial rotation measurement represents the response variable in the experiment.

d. There is reason to believe that there are differences in the medial rotation among the students. Therefore, to control for these difference, measurements will be taken at all three positions for each student (subject). The design described is a randomized block design.

e. The linear model for these data is: $E(y) = \beta_0 + \beta_1 x_1 + \beta_2 x_2 + \beta_3 x_3 + \cdots + \beta_{11} x_{11}$,

where $x_1 = \begin{cases} 1 & \text{if beginning position} \\ 0 & \text{if not} \end{cases}$ $\quad x_2 = \begin{cases} 1 & \text{if change direction} \\ 0 & \text{if not} \end{cases}$

$x_3 = \begin{cases} 1 & \text{if student \#1} \\ 0 & \text{if not} \end{cases}$ $\quad \cdots \quad x_{11} = \begin{cases} 1 & \text{if student \#9} \\ 0 & \text{if not} \end{cases}$

11.21 H_0: $\mu_1 = \mu_2 = \cdots = \mu_7$
H_a: At least two treatment means differ

Test statistic: $F = \dfrac{\text{MST}}{\text{MSE}} = \dfrac{70.22}{1.33} = 52.80$

Rejection region: $\alpha = .05$, $\nu_1 = 6$, $\nu_2 = 54$, $F_{.05} \approx 2.25$
Reject H_0 if $F > 2.25$

Conclusion: Reject H_0. There is sufficient evidence to indicate a difference among mean preference scores for the seven video display combinations at $\alpha = .05$.

11.23 To determine if the presence of plants has an effect on stress levels, we test:

H_0: $\mu_1 = \mu_2 = \mu_3$
H_a: At least two treatment means differ

Test statistic: $F = .019$

The p-value of the test is .981. Since the p-value is greater than any reasonable α, we would not reject H_0. There is insufficient evidence to indicate the presence of plants has an effect on stress levels.

To determine if blocking was effective, we test:

H_0: There is no difference among the block means
H_a: At least two block means differ

Test statistic: $F = .635$

The p-value of the test is .754. Since the p-value is greater than any reasonable α, we would not reject H_0. There is insufficient evidence to indicate that blocking was effective.

11.25 a. The complete model is:

$$E(y) = \beta_0 = \beta_1 x_1 + \beta_2 x_2 + \beta_3 x_3 + \beta_4 x_4 + \beta_5 x_5 + \beta_6 x_6 + \beta_7 x_7 + \beta_8 x_8$$

where $x_1 = \begin{cases} 1 & \text{if Rater 1} \\ 0 & \text{otherwise} \end{cases}$ $x_2 = \begin{cases} 1 & \text{if Rater 2} \\ 0 & \text{otherwise} \end{cases}$

$x_3 = \begin{cases} 1 & \text{if Candidate } A \\ 0 & \text{otherwise} \end{cases}$ $x_4 = \begin{cases} 1 & \text{if Candidate } B \\ 0 & \text{otherwise} \end{cases}$

$x_5 = \begin{cases} 1 & \text{if Candidate } C \\ 0 & \text{otherwise} \end{cases}$ $x_6 = \begin{cases} 1 & \text{if Candidate } D \\ 0 & \text{otherwise} \end{cases}$

$x_7 = \begin{cases} 1 & \text{if Candidate } E \\ 0 & \text{otherwise} \end{cases}$ $x_8 = \begin{cases} 1 & \text{if Candidate } F \\ 0 & \text{otherwise} \end{cases}$

b. The reduced model appropriate to testing for differences in the mean scores given by the three raters is:

$$E(y) = \beta_0 + \beta_3 x_3 + \beta_4 x_4 + \beta_5 x_5 + \beta_6 x_6 + \beta_7 x_7 + \beta_8 x_8$$

c. The reduced model appropriate for determining whether blocking by candidates was effective in removing an unwanted source of variability is:

$$E(y) = \beta_0 + \beta_1 x_1 + \beta_2 x_2$$

11.27 a. H_0: $\mu_1 = \mu_2 = \mu_3$
H_a: At least two machine means differ

Test statistic: $F = \dfrac{\text{MSM}}{\text{MSE}} = 5.06$

Rejection region: $\alpha = .10$, $\nu_1 = 2$, $\nu_2 = 8$, $F_{.10} = 3.11$
Reject H_0 if $F > 3.11$

Conclusion: Reject H_0. There is sufficient evidence (at $\alpha = .10$) to indicate a difference in mean time per call among the three machines.

b. H_0: All function means are equal
H_a: At least two function means differ

Test statistic: $F = \frac{\text{MSF}}{\text{MSE}} = 2.28$

Rejection region: $\alpha = .10$, $\nu_1 = 4$, $\nu_2 = 8$, $F_{.10} = 2.81$
Reject H_0 if $F > 2.81$

Conclusion: Do not reject H_0. There is insufficient evidence (at $\alpha = .10$) to indicate blocking was effective.

11.29 a. The ANOVA table is:

Source	df
Period	1
Gender	1
P*G	1
Error	120
Total	123

b. Complete Model: $E(y) = \beta_0 + \beta_1 x_1 + \beta_2 x_2 + \beta_3 x_1 x_2$, where

$$x_1 = \begin{cases} 1 & \text{if 16 hours or less} \\ 0 & \text{if more than 16 hours} \end{cases} \qquad x_2 = \begin{cases} 1 & \text{if male} \\ 0 & \text{if female} \end{cases}$$

Reduced Model: $E(y) = \beta_0 + \beta_1 x_1 + \beta_2 x_2$

c. Differences in mean body weight gains of males and females do not depend on the photoperiod.

d. Evidence was found in the differences between the two levels of both main effects variables:

$$\mu_{\le 16} \ne \mu_{>16} \text{ and } \mu_M \ne \mu_F$$

11.31 a. The p-value exceeds .01. Since this p-value is not small, H_0 is not rejected. There is insufficient evidence to indicate the two factors interact.

b. The p-value exceeds .01. Since this p-value is not small, H_0 is not rejected. There is insufficient evidence to indicate problem type affects proportion of ideas recalled correctly.

c. The p-value does not exceed .01. Since this p-value is small, H_0 is rejected. There is sufficient evidence to indicate the group affects proportion of ideas recalled correctly.

11.33 a. From the printout, the ANOVA table is:

Source	df	SS	MS	F	p
Amount (A)	3	104.19	34.73	20.12	.0001
Method (B)	3	28.63	9.54	5.53	.0036
AB	9	25.13	2.79	1.62	.1523
Error	32	55.25	1.73		
Total	47	213.20			

b. To determine if interaction exists, we test:

H_0: No interaction between amount and method exists
H_a: Amount and method interact

Test statistic: $F = \dfrac{\text{MS}(AB)}{\text{MSE}} = 1.62$

Rejection region: $\alpha = .01$, $\nu_1 = 9$, $\nu_2 = 32$, $F_{.01} \approx 3.07$
Reject H_0 if $F > 3.07$

Conclusion: Do not reject H_0. There is insufficient evidence to indicate interaction between amount and method exists at $\alpha = .01$.

c. Because interaction is not present, the effects of one independent variable (amount) on the dependent variable (shear strength) is the same at each level of the second independent variable (method) and vice versa.

d. To test for the main effects of factor A, we test:

H_0: There are no differences among the mean shear strengths for the different amounts of antimony
H_a: At least two of the means differ

Test statistic: $F = \dfrac{\text{MS}(A)}{\text{MSE}} = 20.12$

Rejection region: $\alpha = .01$, $\nu_1 = 3$, $\nu_2 = 32$, $F_{.01} \approx 4.51$
Reject H_0 if $F > 4.51$

Conclusion: Reject H_0. There is sufficient evidence to indicate differences in the mean shear strengths among the different amounts of antimony at $\alpha = .01$.

To test for the main effects of factor B, we test:

H_0: There are no differences among the mean shear strengths for the different cooling methods
H_a: At least two of the means differ

Test statistic: $F = \dfrac{\text{MS}(B)}{\text{MSE}} = 5.53$

Rejection region: $\alpha = .01$, $\nu_1 = 3$, $\nu_2 = 32$, $F_{.01} \approx 4.51$
Reject H_0 if $F > 4.51$

Conclusion: Reject H_0. There is sufficient evidence to indicate differences in the mean shear strengths among the different cooling methods at $\alpha = .01$.

11.35 a. There are two factors in this problem with two levels each:

Factor A - Confirmation of accounts receivable, with levels completed and not completed.
Factor B - Verification of sales transactions, with levels completed and not completed.

There are 4 treatments in this problem—the treatments correspond to all possible combinations of the two factors:

Let C_1 = confirmation of accounts receivable completed
C_2 = confirmation of accounts receivable not completed
V_1 = verification of sales transactions completed
V_2 = verification of sales transactions not completed

The four treatments are: C_1V_1, C_1V_2, C_2V_1, C_2V_2

b. If the two factors interact, it means that the effect of Factor A (confirmation of accounts receivable) on account misstatement depends on the level of Factor B (verification of sales transactions).

c. Yes. If interaction does not exist, the two lines drawn would be parallel. In this case, the two lines are not parallel. The difference between the mean accounts misstatements for completed and not completed confirmation of accounts receivable for completed verification of sales transaction is much smaller than the difference between the mean accounts misstatements for completed and not completed confirmation of accounts receivable for not completed verification of sales transaction.

11.37 a. Since the p-value associated with the ANOVA F test for interaction exceeds .10, our conclusion would be:

Do not reject H_0 at $\alpha \leq .10$. There is insufficient evidence to indicate the presence of interaction between window size and jump length.

b. Since the p-value associated with the ANOVA F test for the main effect of window size exceeds .10, our conclusion would be:

Do not reject H_0 at $\alpha \leq .10$. There is insufficient evidence to indicate a difference between the two window sizes.

c. Since the p-value associated with the ANOVA F test for the main effect of jump length is less than .05, our conclusion would be:

Reject H_0 at $\alpha = .05$. There is sufficient evidence to indicate a difference in the mean reading speed among the three jump lengths.

11.39 To determine if interaction exists between inspection levels and burn-in hours, we test:

H_0: No interaction between inspection level and burn-in
H_a: Inspection level and burn-in interact

Test statistic: $F = \dfrac{\text{MS}(AB)}{\text{MSE}} = 95.27$ (from printout)

The p-value is .0001. Since the p-value is so small, H_0 is rejected. There is sufficient evidence to indicate that inspection levels and burn-in hours interact to affect early part failure.

Since interaction exists, no main effect tests are run. The next step is to find which treatment combination gives the optimal detection of early part failure.

11.41 a.

Source	df
A	2
B	3
C	1
AB	6
AC	2
BC	3
ABC	6
Error	120
Total	143

b. H_0: Interactions exist
H_a: At least one interaction exists

Test statistics: $F = \dfrac{\text{MS(Interactions)}}{\text{MSE}} = \dfrac{.73}{.14} = 5.21$

Rejection region: $\alpha = .05$, $\nu_1 = 17$, $\nu_2 = 120$, $F_{.05} \approx 1.75$
Reject H_0 if $F > 1.75$

Conclusion: Reject H_0. There is sufficient evidence to indicate at least one interaction exists among the three factors at $\alpha = .05$.

c. H_0: No interaction between factors A and B
H_a: A and B interact

Test statistic: $F = \dfrac{\text{MS}AB}{\text{MSE}} = \dfrac{.39}{.14} = 2.79$

Rejection region: $\alpha = .05$, $\nu_1 = 6$, $\nu_2 = 120$, $F_{.05} = 2.17$
Reject H_0 if $F > 2.17$

Conclusion: Reject H_0. There is sufficient evidence to indicate AB interaction exists at $\alpha = .05$.

The difference in the mean effect on the whiteness of fine bond paper caused by each of the four bleaches depends on the paper stock being used.

d. SS(Interactions = MS(Interactions) × df(Interactions) = .73(17) = 12.41
SSE = MSE × df(Error) = .14(120) = 16.80

The ANOVA table becomes:

Source	df	SS
A	2	2.35
B	3	2.71
C	1	.72
Interactions	17	12.41
Error	120	16.80
Total	143	34.99

$$\text{SS(Total)} = \text{SS}(A) + \text{SS}(B) + \text{SS}(C) + \text{SS(Interactions)} + \text{SSE}$$
$$= 2.35 + 2.71 + .72 + 12.41 + 16.8 = 34.99$$

$$R^2 = \frac{\text{SS(Total)} - \text{SSE}}{\text{SS(Total)}} = \frac{34.99 - 16.80}{34.99} = .52$$

There is a 52% reduction in the sample variation of the whiteness measures about their mean when we use factors A, B, and C in a $3 \times 4 \times 2$ design.

11.43 a. For cold drawn incoloy alloy, $x_2 = 0$ and $x_3 = 0$.

$$E(y) = \beta_0 + \beta_1 x_1 + \beta_2 x_1^2$$

b. For cold drawn inconel alloy, $x_2 = 1$ and $x_3 = 0$.

$$E(y) = \beta_0 + \beta_1 x_1 + \beta_2 x_1^2 + \beta_3 + \beta_6 x_1 + \beta_9 x_1^2$$
$$= (\beta_0 + \beta_3) + (\beta_1 + \beta_6)x_1 + (\beta_2 + \beta_9)x_1^2$$

c. For cold rolled inconel alloy, $x_2 = 1$ and $x_3 = 1$.

$$E(y) = \beta_0 + \beta_1 x_1 + \beta_2 x_1^2 + \beta_3 + \beta_4 + \beta_5 + \beta_6 x_1 + \beta_7 x_1 + \beta_8 x_1 + \beta_9 x_1^2 + \beta_{10} x_1^2 + \beta_{11} x_1^2$$
$$= (\beta_0 + \beta_3 + \beta_4 + \beta_5) + (\beta_1 + \beta_6 + \beta_7 + \beta_8)x_1 + (\beta_2 + \beta_9 + \beta_{10} + \beta_{11})x_1^2$$

d. The prediction equation is:

$$\hat{y} = 31.15 + .153x_1 - .00396x_1^2 + 17.05x_2 + 19.1x_3 - 14.3x_2x_3 + .151x_1x_2 + .017x_1x_3 - .08x_1x_2x_3 - .00356x_1^2x_2 + .0006x_1^2x_3 + .0012x_1^2x_2x_3$$

e. For inconel alloy and cold rolled, $x_2 = 1$ and $x_3 = 1$.

$$\hat{y} = (31.15 + 17.05 + 19.1 - 14.3) + (.153 + .151 + .017 - .08)x_1 + (-.00396 - .00356 + .0006 + .0012)x_1^2$$
$$= 53 + .241x_1 - .00572x_1^2$$

For incoloy alloy and cold rolled, $x_2 = 0$ and $x_3 = 1$.

$$\hat{y} = (31.15 + 19.1) + (.153 + .017)x_1 + (-.00396 + .0006)x_1^2$$
$$= 50.25 + .17x_1 - .00336x_1^2$$

For inconel alloy and cold drawn, $x_2 = 1$ and $x_3 = 0$.

$$\hat{y} = (31.15 + 17.05) + (.153 + .151)x_1 + (-.00396 + -.00356)x_1^2$$
$$= 48.2 + .304x_1 - .00752x_1^2$$

For incoloy alloy and cold drawn, $x_2 = 0$ and $x_3 = 0$.

$$\hat{y} = 31.15 + .153x_1 - .00396x_1^2$$

f.

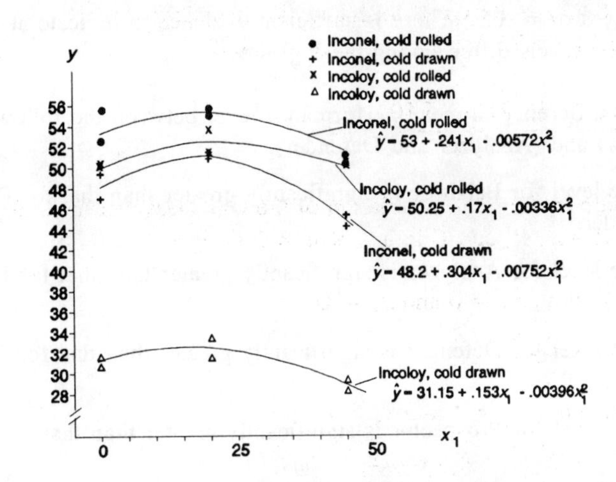

11.45 a. This is a $3 \times 4 \times 3 \times 3$ factorial experiment. The number of different treatments is $3 \times 4 \times 3 \times 3 = 108$.

b. Let $x_1 = \begin{cases} 1 & \text{if control group} \\ 0 & \text{otherwise} \end{cases}$ $x_2 = \begin{cases} 1 & \text{if Guided CAI} \\ 0 & \text{otherwise} \end{cases}$

$x_3 = \begin{cases} 1 & \text{if high school algebra} \\ 0 & \text{otherwise} \end{cases}$ $x_4 = \begin{cases} 1 & \text{if trigonometry} \\ 0 & \text{otherwise} \end{cases}$

$x_5 = \begin{cases} 1 & \text{if differential calculus} \\ 0 & \text{otherwise} \end{cases}$ $x_6 = \begin{cases} 1 & \text{if no computer} \\ 0 & \text{otherwise} \end{cases}$

$x_7 = \begin{cases} 1 & \text{if some computer} \\ 0 & \text{otherwise} \end{cases}$ $x_8 = \begin{cases} 1 & \text{if A in GE 101} \\ 0 & \text{otherwise} \end{cases}$

$x_9 = \begin{cases} 1 & \text{if B in GE 101} \\ 0 & \text{otherwise} \end{cases}$

The complete model will include 108 terms including β_0, 9 main effect terms, 30 two-way interaction terms, 44 three-way interaction terms, and 24 four-way interaction terms.

c. To test to see if interaction exists among the four factors, we test:

H_0: $\beta_{10} = \beta_{11} = \cdots = \beta_{107} = 0$
H_a: At least one of the coefficients is not 0

11.47 a. Rejection region: $\alpha = .05$, $\nu_1 = 5$, $\nu_2 = 366$, $F_{.05} \approx 2.21$
Reject H_0 if $F > 2.21$

Conclusion: Reject H_0. There is sufficient evidence to indicate that at least two of the groups differ in their mean "ad avoidances" at $\alpha = .05$.

b. Using the multiple comparison analysis given, the Time Shifter group has a statistically lower mean "ad avoidance" level than all the other groups. In addition, no other groups can be said to statistically differ in respect to their mean "ad avoidance" levels. Both these statements are made with $\alpha = .05$.

11.49 a. We would test:

H_0: $\mu_R = \mu_D = \mu_P = \mu_A = \mu_B$
H_a: At least two of the five means differ

Conclusion: Reject H_0 at $\alpha = .05$. There is sufficient evidence to indicate at least two of the mean performance levels differ among the 5 groups.

b. There is no significant difference in mean performance level between the following pairs: (Balancer and Analyzer) and (Analyzer and Defender).

The mean performance level for Balancer is significantly greater than that for Reactor, Prospector, and Defender.

The mean performance level for Analyzer is significantly greater than that for Reactor and Prospector.

The mean performance level for Defender is significantly greater than that for Reactor and Prospector.

The mean performance level for Prospector is significantly greater than that for Reactor.

11.51 a. With a p-value $< .0003$, we would reject H_0 at $\alpha > .0003$. There is sufficient evidence to indicate that Accuracy and Vocabulary interact to affect mean completion time.

b. Since there is evidence of interaction between Accuracy and Vocabulary, this means that the effect of Accuracy on the mean completion time depends on the level of Vocabulary. Thus, we need to compare the mean completion times for the 3 levels of Accuracy at each level of Vocabulary.

c. For Vocabulary level 75%:

The mean completion times for the 3 levels of Accuracy are all significantly different

For Vocabulary level 87.5%:

The mean completion times for the 3 levels of Accuracy are all significantly different

For Vocabulary level 100%:

The mean completion times for 90% Accuracy is significantly higher than the mean completion time for Accuracy levels 99% and 95%. There is no significant difference in the mean completion times for Accuracy levels 99% and 95%.

11.53 a. Using $B = .32$ as the critical value, any means which differ by more than .32 are considered significantly different. The following means are different:

2.439 2.683 2.854 3.098 3.293 3.366 3.390 3.561 3.756 3.854 3.878 4.000 4.000 4.049 4.073 4.219 4.293 4.317

The mean rating for Policy 1 is significantly different from all other policy means except the mean of Policy 2.

The mean rating for Policy 2 is significantly different from all other policy means except the means of Policy 1 and Policy 3.

The mean rating for Policy 3 is significantly different from all other policy means except the means of Policy 2 and Policy 4.

The mean rating for Policy 4 is significantly different from all other policy means except the means of Policies 3, 5, 6, and 7.

The mean rating for Policy 5 is significantly different from all other policy means except the means of Policies 4, 6, 7, and 8.

The mean rating for Policy 6 is significantly different from all other policy means except the means of Policies 4, 5, 7, and 8.

The mean rating for Policy 7 is significantly different from all other policy means except the means of Policies 4, 5, 6, and 8.

The mean rating for Policy 8 is significantly different from all other policy means except the means of Policies 5, 6, 7, 9, 10, and 11.

The mean rating for Policy 9 is significantly different from all other policy means except the means of Policies 8, 10, 11, 12, 13, 14, and 15.

The mean rating for Policy 10 is significantly different from all other policy means except the means of Policies 8, 9, 11, 12, 13, 14, and 15.

The mean rating for Policy 11 is significantly different from all other policy means except the means of Policies 8, 9, 10, 12, 13, 14, and 15.

The mean rating for Policy 12 is significantly different from all other policy means except the means of Policies 9, 10, 11, 13, 14, 15, 16, 17, and 18.

The mean rating for Policy 13 is significantly different from all other policy means except the means of Policies 9, 10, 11, 12, 14, 15, 16, 17, and 18.

The mean rating for Policy 14 is significantly different from all other policy means except the means of Policies 9, 10, 11, 12, 13, 15, 16, 17, and 18.

The mean rating for Policy 15 is significantly different from all other policy means except the means of Policies 9, 10, 11, 12, 13, 14, 16, 17, and 18.

The mean rating for Policy 16 is significantly different from all other policy means except the means of Policies 12, 13, 14, 15, 17, and 18.

The mean rating for Policy 17 is significantly different from all other policy means except the means of Policies 12, 13, 14, 15, 16, and 18.

The mean rating for Policy 18 is significantly different from all other policy means except the means of Policies 12, 13, 14, 15, 16, and 17.

b. Yes. There is no significant difference in the mean ratings of policies 12 through 18. All of these policies have sample means of 4.000 or higher.

11.55 From Exercise 11.33, we have $p = 4$ antimony amounts, $\nu = 32$, $n_t = 12$, and $s = \sqrt{\text{MSE}} = \sqrt{1.73} = 1.3153$.

Then for $\alpha = .01$, $q_{.01}(4,32) = 4.80$ and $\omega = q_{.01}(4,32)\frac{s}{\sqrt{n_t}} = 4.80\frac{1.3153}{\sqrt{12}} = 1.823$

The four sample means ranked in order are:

10%	0%	3%	5%
17.017	20.175	20.408	20.617

The means which differ by more than 1.823 (ω) are significantly different. The mean shear strength for 10% antimony is significantly smaller than the mean shear strength for all the other amounts of antimony. There are no other significant differences.

11.57 For $k = 4$ means, the number of pairwise comparisons to be made is:

$$g = \frac{k(k-1)}{2} = \frac{4(3)}{2} = 6$$

Thus, we need to find $t_{\alpha/2g} = t_{.06/2(6)} = t_{.005}$ with $\nu = n - k = 167 - 4 = 163$. From Table 2, Appendix C, $t_{.005} \approx 2.576$.

From Exercise 11.12, MSE = .6092 and $\sqrt{\text{MSE}} = .7805$.

The means arranged in order are:

(1) Jail Correctional	(2) Jail Mental Health	(3) County Mental Health	(4) Other Mental Health
1.81	2.32	2.32	2.84

To compare mean (4) to (1),

$$B_{14} = t_{.005}(s)\sqrt{\frac{1}{n_1} + \frac{1}{n_4}} = 2.576(.7805)\sqrt{\frac{1}{58} + \frac{1}{28}} = .4627$$

Since $\bar{x}_4 - \bar{x}_1 = 2.84 - 1.81 = 1.03 > .4627$, means 4 and 1 are significantly different.

To compare (4) to (2),

$$B_{24} = 2.576(.7805)\sqrt{\frac{1}{29} + \frac{1}{28}} = .5327$$

Since $\bar{x}_4 - \bar{x}_2 = 2.84 - 2.32 = .52 < .5327$, there is no significant difference between means 2 and 4.

Since means 2 and 4 are connected, we do not need to compare means 2 and 3 and means 3 and 4. They are not significantly different.

To compare (3) to (1),

$$B_{13} = 2.576(.7805)\sqrt{\frac{1}{58} + \frac{1}{52}} = .3840$$

Since $\bar{x}_3 - \bar{x}_1 = 2.32 - 1.81 = .51 > .3840$, means 3 and 1 are significantly different.

To compare (2) to (1),

$$B_{12} = 2.576(.7805)\sqrt{\frac{1}{58} + \frac{1}{29}} = .4573$$

Since $\bar{x}_2 - \bar{x}_1 = 2.32 - 1.81 = .51 > .4573$, means 2 and 1 are significantly different.

Thus, the mean level of conflict perceived by jail correctional staff is significantly less than the means of the other three groups. There are no other significant differences.

11.59 a. First we look for any differences in the mean rate of penetration for the three bits. We test:

H_0: $\mu_1 = \mu_2 = \mu_3$
H_a: At least two means differ

The test statistic is $F = 9.50$.

The p-value is $p = .006$.

For $\alpha = .01$, we can reject H_0 and conclude that at least two bits have different mean penetration rates.

b. From the exercise, $p = 3$, $\nu = \text{df(error)} = 9$, and MSE $= 19.3$. There are $n_t = 4$ observations per treatment.

Then for $\alpha = .05$, $q_{.05}(3,9) = 3.95$ and $\omega = q_{.05}(3,9)\dfrac{s}{\sqrt{n_t}} = 3.95\dfrac{\sqrt{19.3}}{\sqrt{4}} = 8.677$

The three sample means ranked in order are:

IADC 5-1-7	IADC 1-2-6	PD-1
20.775	28.050	34.300

(Lines beneath: one under IADC 1-2-6 and PD-1; one under IADC 5-1-7 and IADC 1-2-6.)

The sample means that differ by more than 8.677 are significantly different. The mean speed for PD-1 is significantly faster than the mean speed for IADC 5-1-7.

c. From this exercise, we have $p = 3$ treatments, $\nu = \text{df(error)} = 9$, and MSE $= 19.3$. There are $n_i = n_j = 4$, observations per treatment for all pairs of treatments.

Then for $\alpha = .05$, $F_{.05}(2, 9) = 4.26$ and

$$S_{ij} = \sqrt{(p-1)F_{.05}(2, 9)\text{MSE}\left(\frac{1}{n_i} + \frac{1}{n_j}\right)} = \sqrt{(2)(4.26)(19.3)\left(\frac{1}{4} + \frac{1}{4}\right)} = 9.07$$

The 3 sample means ranked in order are:

IADC 5-1-7	IADC 1-2-6	PD-1
20.775	28.05	34.3

(Lines beneath: one under IADC 1-2-6 and PD-1; one under IADC 5-1-7 and IADC 1-2-6.)

The pairs of means that are significantly different are:

(PD-1, IADC 5-1-7)

d. For this exercise, we have $p = 3$, $s = \sqrt{\text{MSE}} = \sqrt{19.3} = 4.393$, $\nu = 9$, and $n_i = n_j = 4$ for all treatment pairs (i, j). For $p = 3$ means, the number of pairwise comparisons to be made is:

$$g = \frac{p(p-1)}{2} = \frac{3(2)}{2} = 3$$

We need to find $t_{\alpha/2g} = t_{.05/2(3)} = t_{.0083}$ for the t distribution based on $\nu = 9$ df. This value is approximately 2.967, and

$$B_{ij} \approx t_{.0083} s \sqrt{\frac{1}{n_i} + \frac{1}{n_j}} = 2.967(4.393)\sqrt{\frac{1}{4} + \frac{1}{4}} = 9.216$$

The 3 sample means ranked in order are:

IADC 5-1-7	IADC 1-2-6	PD-1
20.775	28.05	34.3

The pairs of means that are significantly different are:

(PD-1, IADC 5-1-7)

e. Although the critical value for Tukey's comparison is smaller than that for Scheffé's and Bonferroni's, the conclusions are the same.

11.61 First, we must compute the residuals for each treatment. The residuals are computed by finding the sample mean for each treatment and then subtracting this sample mean from each observation. The sample means for the four treatments are:

Oyster Tissue: $\bar{y}_1 = 1.33$ Citrus Leaves: $\bar{y}_2 = 3.16$
Bovine Liver: $\bar{y}_3 = 0.41$ Human Serum: $\bar{y}_4 = 0.146$

The residuals for each of the treatments are:

Oyster Tissue: 1.02, −.03, −.99 Citrus Leaves: −.84, −.09, .93
Bovine Liver: −.02, .13, −.11 Human Serum: −.046, .024, −.006, .014, .014

Since there are so few residuals per treatment, histograms to check for normality will not be of much use. Also, the analysis of variance is fairly robust with respect to normality. That is, the analysis of variance is valid even if the data are not exactly normally distributed.

We will construct a residual frequency plot to check for differences in variances among the four treatments:

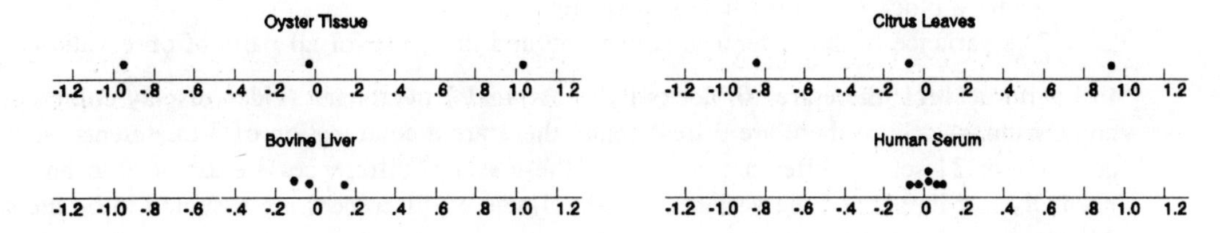

From the above plots, it appears that the variances are not equal. The spreads of the residuals for Oyster Tissue and Citrus Leaves are much greater than the spreads of the residuals for Bovine Liver and Human Serum.

Bartlett's test for the equality of variances is as follows:

First, we must compute the sample variances for each of the four treatments:

Oyster Tissue: $s_1^2 = 1.0107$ Citrus Leaves: $s_2^2 = 0.7893$
Bovine Liver: $s_3^2 = 0.0147$ Human Serum: $s_4^2 = 0.00078$

$$\bar{s}^2 = \frac{\sum (n_i - 1)s_i^2}{\sum (n_i - 1)} = \frac{(3-1)1.0107 + (3-1).7893 + (3-1).0147 + (5-1).00078}{(3-1) + (3-1) + (3-1) + (5-1)}$$
$$= .363252$$

The test is as follows:

H_0: $\sigma_1^2 = \sigma_2^2 = \sigma_3^2 = \sigma_4^2$
H_a: At least two variances differ

$$\text{Test statistic: } B = \frac{\left[\sum (n_i - 1)\right]\ln \bar{s}^2 - \sum (n_i - 1)\ln s_i^2}{1 + \frac{1}{3(p-1)}\left\{\sum \frac{1}{(n_i - 1)} - \frac{1}{\sum (n_i - 1)}\right\}}$$

$$= \frac{10\ln(.363252) - [(3-1)\ln(1.0107) + (3-1)\ln(.7893) + (3-1)\ln(.0147) + (5-1)\ln(.00078)]}{1 + \frac{1}{3(4-1)}\left\{\frac{1}{(3-1)} + \frac{1}{(3-1)} + \frac{1}{(5-1)} - \frac{1}{10}\right\}}$$
$$= 23.1465$$

Rejection region: $\alpha = .05$, df $= p - 1 = 3$, $\chi^2_{.05} = 7.81473$
Reject H_0 if $B > 7.81473$

Conclusion: Reject H_0. There is sufficient evidence to indicate at least two of the variances are different at $\alpha = .05$.

Thus, the analysis may not be valid since the assumption of equal variances is not met.

11.63 The assumptions for the randomized block design are:

1. The probability distribution of the difference between any pair of treatment observations within a block is approximately normal.
2. The variance of the difference is constant and the same for all pairs of observations.

In Exercise 11.21, there are 10 blocks (students) and 7 treatments (video display color combinations). Since there are 7 treatments, there are a combination of 7 treatments taken two at a time or 21 sets of differences. Each of these sets of differences are assumed to be normally distributed and the variances of all 21 sets of differences are assumed to be the same. The 21 sets of differences are:

GB-WB	GB-YW	GB-OW	GB-Y	GB-YA	GB-YO	WB-YW	WB-OW	WB-Y	WB-YA	WB-YO	YW-OW	YW-Y	YW-YA	YW-YO	OW-Y	OW-YA	OW-YO	Y-YA	Y-YO	YA-YO
1	0	5	−1	−2	4	−1	4	−2	−3	3	5	−1	−2	4	−6	−7	−1	−1	5	6
2	−1	4	−1	0	7	−3	2	−3	−2	5	5	0	1	8	−5	−4	3	1	8	7
0	−2	4	−1	−3	3	−2	4	−1	−3	3	6	1	−1	5	−5	−7	−1	−2	4	6
−1	1	3	1	−3	3	2	4	2	−2	4	2	0	−4	2	−2	−6	0	−4	2	6
1	1	6	0	0	7	0	5	−1	−1	6	5	−1	−1	6	−6	−6	1	0	7	7
2	1	5	0	0	6	−1	3	−2	−2	4	4	−1	−1	5	−5	−5	1	0	6	6
−1	−2	2	0	−3	1	−1	3	1	−2	2	4	2	−1	3	−2	−5	−1	−3	1	4
1	−2	5	−2	−3	5	−3	4	−3	−4	4	7	0	−1	7	−7	−8	0	−1	7	8
0	1	7	0	1	9	1	7	0	1	9	6	−1	0	8	−7	−6	2	1	9	8
1	1	6	0	−1	8	0	5	−1	−2	7	5	−1	−2	7	−6	−7	2	−1	8	9

Since there are so few observations (10) per set of differences, histograms to check for normality will not be of much use. Also, analysis of variance is fairly robust with respect to normality. That is, the analysis of variance is valid even if the data are not exactly normally distributed.

Bartlett's test for the equality of variances is as follows:

First, we must compute the sample variances for each set of differences. The 21 sample variances are:

1.16, 1.96, 2.23, 0.71, 2.49, 6.46, 2.62, 1.88, 2.67, 1.78, 4.46, 1.88, 1.07, 1.73, 4.28, 3.21, 1.43, 2.04, 2.67, 7.12, 2.01

$$\bar{s}^2 = \frac{\sum s_i^2}{p} = \frac{1.16 + 1.96 + 2.23 + .71 + \cdots + 2.01}{21} = 2.66$$

The test is as follows:

H_0: $\sigma_1^2 = \sigma_2^2 = \sigma_3^2 = \cdots = \sigma_{21}^2$

H_a: At least two variances differ

Test statistic: $$B = \frac{(n - 1)\left[p \ln \bar{s}^2 - \sum \ln s_i^2\right]}{1 + \frac{(p + 1)}{3p(n - 1)}}$$

$$= \frac{(10 - 1)[21 \ln(2.66) - \ln(1.16) - \ln(1.96) - \ln(2.23) - \cdots - \ln(2.01)]}{1 + \frac{21 + 1}{3(21)(10 - 1)}} = 28.216$$

Rejection region: $\alpha = .05$, df $= p - 1 = 20$, $\chi^2_{.05} = 31.4104$
Reject H_0 if $B > 31.4104$

Conclusion: Do not reject H_0. There is insufficient evidence to indicate at least two of the variances are different at $\alpha = .05$.

Thus, the assumptions are probably valid.

11.65 First, we must compute the residuals for each factor level combination. The residuals are computed by finding the sample mean for each treatment and then subtracting this sample mean from each observation. The sample means for the 27 treatments are:

Burn-in (hours)	Full Military Specification, A	Inspection Levels Reduced Military Specification, B	Commercial, C
1	7.59	7.33	6.17
2	6.95	6.05	5.78
3	6.25	5.57	5.40
4	5.67	5.31	5.66
5	5.22	4.94	5.93
6	4.00	4.52	6.65
7	3.93	3.90	7.69
8	3.75	4.23	8.30
9	3.49	5.38	9.37

The residuals are found by subtracting the above sample means from the observations in each factor level combination. Since there are so few residuals per factor level combination, histograms to check for normality will not be of much use. Also, analysis of variance is fairly robust with respect to normality. That is, the analysis of variance is valid even if the data are not exactly normally distributed.

To check for equal variances, we must compute the sample variances for each factor level combination. The variances are:

Burn-in (hours)	Full Military Specification, A	Inspection Levels Reduced Military Specification, B	Commercial, C
1	.0073	.1033	.0016
2	.2201	.0300	.1414
3	.1276	.0633	.0025
4	.0931	.0034	.0344
5	.0399	.0061	.0658
6	.1200	.0012	.0097
7	.0785	.0042	.0463
8	.0037	.1047	.1300
9	.0030	.0481	.2385

$$\bar{s}^2 = \frac{\sum s_i^2}{p} = \frac{.0073 + .2201 + .1276 + .0931 + \cdots + .2385}{27} = .063988$$

The test is as follows:

H_0: $\sigma_1^2 = \sigma_2^2 = \sigma_3^2 = \cdots = \sigma_{27}^2$
H_a: At least two variances differ

Test statistic: $B = \dfrac{(n-1)\left[p \ln \bar{s}^2 - \sum \ln s_i^2\right]}{1 + \dfrac{(p+1)}{3p(n-1)}}$

$$= \frac{(3-1)[27 \ln(.063988) - \ln(.0073) - \ln(.2201) - \cdots - \ln(.2385)]}{1 + \dfrac{27+1}{3(27)(3-1)}} = 41.62$$

Rejection region: $\alpha = .05$, df $= p - 1 = 26$, $\chi^2_{.05} = 38.8852$
Reject H_0 if $B > 38.8852$

Conclusion: Reject H_0. There is sufficient evidence to indicate at least two of the variances are different at $\alpha = .05$.

Thus, the assumption of equal variances is probably not valid. Note: If $\alpha = .01$ ($\chi^2_{.01} = 45.6417$), H_0 would not be rejected.

11.67 a. The three treatments are the three stores: Winn-Dixie, Publix, and Kash N'Karry. The blocks are the 60 grocery items.

b. $E(y) = \beta_0 + \beta_1 x_1 + \beta_2 x_2 + \beta_3 x_3 + \beta_4 x_4 + \cdots + \beta_{61} x_{61}$

$x_1 = \begin{cases} 1 \text{ if Winn-Dixie} \\ 0 \text{ if not} \end{cases}$ $\quad x_2 = \begin{cases} 1 \text{ if Publix} \\ 0 \text{ if not} \end{cases}$

$x_3 = \begin{cases} 1 \text{ if Big Thirst Towel} \\ 0 \text{ if not} \end{cases}$ $\quad x_4 = \begin{cases} 1 \text{ if Camp Crm/Broccoli} \\ 0 \text{ if not} \end{cases}$

$\cdots$ $\quad x_{61} = \begin{cases} 1 \text{ if Gillette Atra Plus} \\ 0 \text{ if not} \end{cases}$

c. $E(y) = \beta_0 + \beta_3 x_3 + \beta_4 x_4 + \cdots + \beta_{61} x_{61}$, where x_3 x_4, $\cdots$ x_{61} are defined above.

d. The following printouts are generated by fitting the two models using SAS:

```
Model1
Dependent Variable:  Y

                      Analysis of Variance

                        Sum of        Mean
Source          DF     Squares      Square      F Value     Prob > F

Model           61   218.23620     3.57764      106.270       0.0001
Error          118     3.97253     0.03367
C Total        179   222.20873

      Root MSE        0.18346    R-square        0.9821
      Dep Mean        1.83678    Adj R-sq        0.9729
      C.V.            9.98932
```

Parameter Estimates

Variable	DF	Parameter Estimate	Standard Error	T for H0: Parameter=0	Prob > ¦T¦
INTERCEP	1	1.165222	0.10768428	10.821	0.0001
X1	1	-0.259833	0.03349902	-7.756	0.0001
X2	1	-0.005833	0.03349902	-0.174	0.8621
X3	1	0.353333	0.14981217	2.359	0.0200
X4	1	-0.446667	0.14981217	-2.982	0.0035
X5	1	2.246667	0.14981217	14.997	0.0001
X6	1	0.246667	0.14981217	1.647	0.1023
X7	1	0.580000	0.14981217	3.872	0.0002
X8	1	1.013333	0.14981217	6.764	0.0001
X9	1	0.030000	0.14981217	0.200	0.8416
X10	1	1.963333	0.14981217	13.105	0.0001
X11	1	0.946667	0.14981217	6.319	0.0001
X12	1	-0.186667	0.14981217	-1.246	0.2152
X13	1	0.406667	0.14981217	2.715	0.0076
X14	1	0.173333	0.14981217	1.157	0.2496
X15	1	0.580000	0.14981217	3.872	0.0002
X16	1	1.046667	0.14981217	6.987	0.0001
X17	1	-0.016667	0.14981217	-0.111	0.9116
X18	1	0.576667	0.14981217	3.849	0.0002
X19	1	0.480000	0.14981217	3.204	0.0017
X20	1	1.410000	0.14981217	9.412	0.0001
X21	1	-0.113333	0.14981217	-0.757	0.4509
X22	1	-0.320000	0.14981217	-2.136	0.0347
X23	1	0.970000	0.14981217	6.475	0.0001
X24	1	0.066667	0.14981217	0.445	0.6571
X25	1	0.013333	0.14981217	0.089	0.9292
X26	1	0.040000	0.14981217	0.267	0.7899
X27	1	0.213333	0.14981217	1.424	0.1571
X28	1	-0.450000	0.14981217	-3.004	0.0033
X29	1	1.413333	0.14981217	9.434	0.0001
X30	1	2.066667	0.14981217	13.795	0.0001
X31	1	2.500000	0.14981217	16.688	0.0001
X32	1	1.623333	0.14981217	10.836	0.0001
X33	1	0.043333	0.14981217	0.289	0.7729
X34	1	1.176667	0.14981217	7.854	0.0001
X35	1	-0.063333	0.14981217	-0.423	0.6732
X36	1	1.060000	0.14981217	7.076	0.0001
X37	1	0.746667	0.14981217	4.984	0.0001
X38	1	-0.256667	0.14981217	-1.713	0.0893
X39	1	0.213333	0.14981217	1.424	0.1571
X40	1	0.710000	0.14981217	4.739	0.0001
X41	1	-0.573333	0.14981217	-3.827	0.0002
X42	1	0.060000	0.14981217	0.401	0.6895
X43	1	0.086667	0.14981217	0.579	0.5640
X44	1	-0.286667	0.14981217	-1.914	0.0581
X45	1	-0.186667	0.14981217	-1.246	0.2152
X46	1	0.043333	0.14981217	0.289	0.7729
X47	1	0.083333	0.14981217	0.556	0.5791
X48	1	1.426667	0.14981217	9.523	0.0001
X49	1	-0.613333	0.14981217	-4.094	0.0001
X50	1	0.410000	0.14981217	2.737	0.0072
X51	1	1.780000	0.14981217	11.882	0.0001
X52	1	0.090000	0.14981217	0.601	0.5492
X53	1	2.650000	0.14981217	17.689	0.0001
X54	1	2.513333	0.14981217	16.777	0.0001
X55	1	4.673333	0.14981217	31.195	0.0001
X56	1	1.113333	0.14981217	7.432	0.0001
X57	1	1.773333	0.14981217	11.837	0.0001
X58	1	2.593333	0.14981217	17.311	0.0001
X59	1	0.683333	0.14981217	4.561	0.0001
X60	1	0.030000	0.14981217	0.200	0.8416
X61	1	4.200000	0.14981217	28.035	0.0001

Model2
Dependent Variable: Y

Analysis of Variance

Source	DF	Sum of Squares	Mean Square	F Value	Prob > F
Model	59	215.59493	3.65415	66.300	0.0001
Error	120	6.61380	0.05511		
C Total	179	222.20873			

Root MSE	0.23477	R-square	0.9702
Dep Mean	1.83678	Adj R-sq	0.9556
C.V.	12.78140		

Parameter Estimates

Variable	DF	Parameter Estimate	Standard Error	T for H0: Parameter=0	Prob > ¦T¦
INTERCEP	1	1.076667	0.13554212	7.943	0.0001
X3	1	0.353333	0.19168551	1.843	0.0678
X4	1	-0.446667	0.19168551	-2.330	0.0215
X5	1	2.246667	0.19168551	11.721	0.0001
X6	1	0.246667	0.19168551	1.287	0.2006
X7	1	0.580000	0.19168551	3.026	0.0030
X8	1	1.013333	0.19168551	5.286	0.0001
X9	1	0.030000	0.19168551	0.157	0.8759
X10	1	1.963333	0.19168551	10.242	0.0001
X11	1	0.946667	0.19168551	4.939	0.0001
X12	1	-0.186667	0.19168551	-0.974	0.3321
X13	1	0.406667	0.19168551	2.122	0.0359
X14	1	0.173333	0.19168551	0.904	0.3677
X15	1	0.580000	0.19168551	3.026	0.0030
X16	1	1.046667	0.19168551	5.460	0.0001
X17	1	-0.016667	0.19168551	-0.087	0.9309
X18	1	0.576667	0.19168551	3.008	0.0032
X19	1	0.480000	0.19168551	2.504	0.0136
X20	1	1.410000	0.19168551	7.356	0.0001
X21	1	-0.113333	0.19168551	-0.591	0.5555
X22	1	-0.320000	0.19168551	-1.669	0.0976
X23	1	0.970000	0.19168551	5.060	0.0001
X24	1	0.066667	0.19168551	0.348	0.7286
X25	1	0.013333	0.19168551	0.070	0.9447
X26	1	0.040000	0.19168551	0.209	0.8351
X27	1	0.213333	0.19168551	1.113	0.2680
X28	1	-0.450000	0.19168551	-2.348	0.0205
X29	1	1.413333	0.19168551	7.373	0.0001
X30	1	2.066667	0.19168551	10.782	0.0001
X31	1	2.500000	0.19168551	13.042	0.0001
X32	1	1.623333	0.19168551	8.469	0.0001
X33	1	0.043333	0.19168551	0.226	0.8215
X34	1	1.176667	0.19168551	6.139	0.0001
X35	1	-0.063333	0.19168551	-0.330	0.7417
X36	1	1.060000	0.19168551	5.530	0.0001
X37	1	0.746667	0.19168551	3.895	0.0002
X38	1	-0.256667	0.19168551	-1.339	0.1831
X39	1	0.213333	0.19168551	1.113	0.2680
X40	1	0.710000	0.19168551	3.704	0.0003
X41	1	-0.573333	0.19168551	-2.991	0.0034
X42	1	0.060000	0.19168551	0.313	0.7548
X43	1	0.086667	0.19168551	0.452	0.6520
X44	1	-0.286667	0.19168551	-1.496	0.1374
X45	1	-0.186667	0.19168551	-0.974	0.3321
X46	1	0.043333	0.19168551	0.226	0.8215
X47	1	0.083333	0.19168551	0.435	0.6645
X48	1	1.426667	0.19168551	7.443	0.0001
X49	1	-0.613333	0.19168551	-3.200	0.0018
X50	1	0.410000	0.19168551	2.139	0.0345
X51	1	1.780000	0.19168551	9.286	0.0001
X52	1	0.090000	0.19168551	0.470	0.6396
X53	1	2.650000	0.19168551	13.825	0.0001
X54	1	2.513333	0.19168551	13.112	0.0001
X55	1	4.673333	0.19168551	24.380	0.0001

X56	1	1.113333	0.19168551	5.808	0.0001
X57	1	1.773333	0.19168551	9.251	0.0001
X58	1	2.593333	0.19168551	13.529	0.0001
X59	1	0.683333	0.19168551	3.565	0.0005
X60	1	0.030000	0.19168551	0.157	0.8759
X61	1	4.200000	0.19168551	21.911	0.0001

To determine if there is a difference in mean price among the three stores, we test:

H_0: $\beta_1 = \beta_2 = 0$
H_a: At least one of the β_i's $\neq 0$ $i = 1, 2$

Test statistic: $F = \dfrac{(\text{SSE}_R - \text{SSE}_C)/2}{\text{SSE}_C/[180 - (61 + 1)]} = \dfrac{(6.6138 - 3.97253)/2}{3.97253/118} = 39.23$

Rejection region: $\alpha = .05$, $\nu_1 = 2$, $\nu_2 = 118$, $F_{.05} \approx 3.07$
Reject H_0 if $F > 3.07$

Conclusion: Reject H_0. There is sufficient evidence to indicate there is a difference in mean price among the three stores at $\alpha = .05$.

e. The ANOVA table is:

Source	df	SS	MS	F	p
Supermarket	2	2.6413	1.3206	39.23	.0001
Items	59	215.5949	3.6542	108.54	.0001
Error	118	3.9725	0.0337		
Total	179	222.2087			

Yes, the F in this part, 39.23, is the same as in part (d).

f. H_0: Block means are the same
H_a: At least two block means differ

Test statistic: $F = 108.54$

The p-value is .0001. Since the p-value is so small, H_0 is rejected. There is sufficient evidence to indicate that blocking on grocery items was effective in reducing an extraneous source of variation.

g. For this exercise, $p = 3$ grocery stores, $\nu = 118$, $n_t = 60$, and $s = \sqrt{\text{MSE}} = .1835$

Then for $\alpha = .01$, $q_{.01}(3,118) \approx 4.20$ and $\omega = q_{.01}(3,118)\dfrac{s}{\sqrt{n_t}} = 4.20\dfrac{.1835}{\sqrt{60}}$

$= .0995$

The sample means ranked in order are:

Winn-Dixie	Publix	Kash N'Karry
1.666	1.920	1.925

The sample means that differ by more than .0995 are significantly different. The mean price for Winn-Dixie is significantly less than the mean price for either Publix or Kash N'Karry. No other means are significantly different.

11.69 First, we summarize the sample data.

LOCATION				
I	II	III	IV	V
$n_1 = 10$	$n_2 = 10$	$n_3 = 10$	$n_4 = 8$	$n_5 = 10$
$T_1 = 3409$	$T_2 = 4627$	$T_3 = 2861$	$T_4 = 4274$	$T_5 = 3749$
$\bar{T}_1 = 340.9$	$\bar{T}_2 = 462.7$	$\bar{T}_3 = 286.1$	$\bar{T}_4 = 534.25$	$\bar{T}_5 = 374.9$

Also, $\sum y = 18{,}920$ and $\sum y^2 = 8{,}025{,}206$. To test:

H_0: $\mu_1 = \mu_2 = \mu_3 = \mu_4 = \mu_5$

versus H_a: At least two of the means differ

where μ_1 represents the average daily traffic density at location 1 and μ_2, μ_3, μ_4, and μ_5 are similarly defined for locations 2, 3, 4, and 5.

Some preliminary calculations are:

$$\text{CM} = \frac{(\sum y)^2}{n} = \frac{(18{,}920)^2}{48} = 7{,}457{,}633.33$$

$$\text{SS(Total)} = \sum y^2 - \text{CM} = 8{,}025{,}206 - 7{,}457{,}633.3 = 567{,}572.67$$

$$\text{SST} = \frac{T_1^2}{n_1} + \cdots + \frac{T_5^2}{n_5} - \text{CM}$$

$$= \frac{(3409)^2}{10} + \frac{(4627)^2}{10} + \frac{(2861)^2}{10} + \frac{(4274)^2}{8} + \frac{(3749)^2}{10} - 7{,}457{,}633.3$$

$$= 7{,}810{,}457.7 - 7{,}457{,}633.33 = 352{,}824.43$$

$$\text{SSE} = \text{SS(Total)} - \text{SST} = 567{,}572.67 - 352{,}824.43 = 214{,}748.24$$

$$\text{MST} = \frac{\text{SST}}{p-1} = \frac{352{,}824.43}{4} = 88{,}206.108$$

$$\text{MSE} = \frac{\text{SSE}}{n-p} = \frac{214{,}748.24}{43} = 4{,}994.145$$

The complete test is:

H_0: $\mu_1 = \mu_2 = \mu_3 = \mu_4 = \mu_5$
H_a: At least two of the means differ

Test statistic: $F = \frac{\text{MST}}{\text{MSE}} = \frac{88{,}206.108}{4994.145} = 17.66$

Rejection region: $\alpha = .10$, $\nu_1 = 4$, $\nu_2 = 43$, $F_{.10} = 2.09$
Reject H_0 if $F > 2.09$

Conclusion: Reject H_0 at $\alpha = .10$. There is sufficient evidence to indicate a difference in average daily traffic density among the five locations.

Because there is evidence of differences among the treatment means, we will run a multiple comparison using Scheffe's method. We have MSE $= 4994.145$, $p = 5$, and $\nu =$ df(Error) $= 43$. Also, $n_1 = n_2 = n_3 = n_5 = 10$ and $n_4 = 8$. There will be 2 critical values:

For $\alpha = .10$, $F_{.10}(4, 43) \approx 2.09$ and

$$S_{4j} = \sqrt{(p-1)F_{.10}(4, 43)\text{MSE}\left(\frac{1}{n_4} + \frac{1}{n_j}\right)} \quad \textit{for } j \neq 4$$

$$= \sqrt{(4)(2.09)(4994.145)\left(\frac{1}{8} + \frac{1}{10}\right)} = 96.92$$

$$S_{ij} = \sqrt{(p-1)F_{.10}(4, 43)\text{MSE}\left(\frac{1}{n_i} + \frac{1}{n_j}\right)} \quad \textit{for } i \neq 4,\ j \neq 4$$

$$= \sqrt{(4)(2.09)(4994.145)\left(\frac{1}{10} + \frac{1}{10}\right)} = 91.38$$

The sample means arranged in order are:

III	I	V	II	IV
286.1	340.9	374.9	462.7	534.25

There are significant differences between the following pairs of means: (IV, V), (IV, I), (IV, III), (II, I) and (II, III)

11.71 a. First, we test for interaction between nationality and industry.

To determine if nationality and industry interact to affect the mean degree of humor, we test:

H_0: Nationality and industry do not interact
H_a: Nationality and industry do interact

The test statistic is $F = 1.38$.

The p-value associated with this test statistic is .046. Since this p-value is less than $\alpha = .05$, H_0 is rejected. There is sufficient evidence to indicate that nationality and industry interact to affect mean degree of humor at $\alpha = .05$.

Since interaction is significant, the main effect tests are not run.

b. These conclusions agree with the conclusions reached in part (a). The effect of nationality on the mean degree of humor depends on the industry being considered. This indicates that interaction is present.

11.73 Since overbars connect means that are not significantly different, no difference can be detected between the following treatment pairs:

TRT	Pairs
Rest	*A*
C	*D*
C	*E*
D	*E*

There is a significant difference between the following treatment pairs:

TRT	Pairs	TRT	Pairs
Rest	*B*	*A*	*D*
Rest	*C*	*A*	*E*
Rest	*D*	*B*	*C*
Rest	*E*	*B*	*D*
A	*B*	*B*	*E*
A	*C*		

11.75 a. A 3 × 5 factorial design was employed.

b. Since there is no replication, the degrees of freedom for error will be 0.

c. $E(y) = \beta_0 + \beta_1 x_1 + \beta_2 x_2 + \beta_3 x_1 x_2 + \beta_4 x_1^2 + \beta_5 x_2^2$

d. To test for interaction we would test H_0: $\beta_3 = 0$ by proposing the reduced model:

$$E(y) = \beta_0 + \beta_1 x_1 + \beta_2 x_2 + \beta_4 x_1^2 + \beta_5 x_2^2$$

e. The printout from fitting the complete second order model using SAS is:

```
Model1
Dependent Variable:  Y

                        Analysis of Variance

Source          DF      Sum of          Mean      F Value    Prob > F
                       Squares        Square

Model            5   554.64305     110.92861       13.890      0.0005
Error            9    71.87428       7.98603
C Total         14   626.51733

      Root MSE       2.82596   R-square        0.8853
      Dep Mean      12.38667   Adj R-sq        0.8215
      C.V.          22.81451

                        Parameter Estimates

                      Parameter       Standard     T for H0:
Variable    DF         Estimate          Error   Parameter=0    Prob > |T|

INTERCEP     1      -384.749461   127.80093139        -3.011        0.0147
X1           1         3.728461     1.34279582         2.777        0.0215
X2           1        12.724177     5.13634031         2.477        0.0351
X1SQ         1        -0.008711     0.00350276        -2.487        0.0346
X2SQ         1        -0.321944     0.34181269        -0.942        0.3709
X1X2         1        -0.049862     0.02437619        -2.046        0.0711
```

The fitted model is:

$$\hat{y} = -384.749 + 3.728x_1 + 12.724x_2 - .0087x_1^2 - .3219x_2^2 - 0.499x_1x_2$$

To determine if interaction is present, we test:

H_0: $\beta_5 = 0$
H_a: $\beta_5 \neq 0$

Test statistic: $t = -2.046$

Conclusion: The p-value is .0711. Do not reject H_0. There is insufficient evidence to indicate that interaction is present between time and temperature at $\alpha = .05$.

11.77 a. The appropriate linear model for this study is:

$$E(y) = \beta_0 = \beta_1 x_1 + \beta_2 x_2$$

where $x_1 = \begin{cases} 1 & \text{if small company} \\ 0 & \text{otherwise} \end{cases}$ $x_2 = \begin{cases} 1 & \text{if medium company} \\ 0 & \text{otherwise} \end{cases}$

b. The hypotheses are:

H_0: $\mu_1 = \mu_2 = \mu_3$
H_a: At least two of the three treatment means differ

or

H_0: $\beta_1 = \beta_2 = 0$
H_a: At least one $\beta_i \neq 0$

c. df (Company size) $= p - 1 = 3 - 1 = 2$
df (Error) $= n - \text{p} = 124 - 3 = 121$
MSE = SSE/df = 190/121 = 1.5702

We know F = MST/MSE ⇒ MST = F(MSE) = 4.93(1.5702) = 7.7411
We know MST = SST/df ⇒ SST = df(MST) = 2(7.7411) = 15.4822

The ANOVA table is:

Source	df	SS	MS	F
Company Size	2	15.4822	7.7411	4.93
Error	121	190	1.5702	
Total	123	205.4822		

d. H_0: $\mu_1 = \mu_2 = \mu_3$
H_a: At least two means differ

Test statistic: $F = 4.93$

Rejection region: $\alpha = .05$, $\nu_1 = 2$, $\nu_2 = 121$, $F_{.05} \approx 3.07$
Reject H_0 if $F > 3.07$

Conclusion: Reject H_0. There is sufficient evidence to indicate the mean employability ratings of physically handicapped persons differ among the three groups at $\alpha = .05$.

11.79 a.

Source	df	SS	MS	F
Method	2	0.19	0.10	.08
Brand	5	605.70	121.14	93.18
Error	10	13.05	1.30	
Total	17	618.94		

b. **Method (Treatment) Totals**

$T_1 = 57.4 \qquad T_2 = 58.8 \qquad T_3 = 58.6$

Brand (Block) Totals

$B_1 = 57.7 \qquad B_3 = 27.5 \qquad B_5 = 16.0$
$B_2 = 46.6 \qquad B_4 = 17.5 \qquad B_6 = 9.5$

$\sum y = 174.8, \sum y^2 = 2316.44$

$$\text{CM} = \frac{(\sum y)^2}{n} = \frac{(174.8)^2}{18} = 1697.5022$$

$$\text{SS(Total)} = \sum y^2 - \text{CM} = 618.9378$$

$$\text{SST} = \frac{57.4^2}{6} + \frac{58.8^2}{6} + \frac{58.6^2}{6} - \text{CM} = 1697.6933 - \text{CM} = .1911$$

$$\text{SSB} = \frac{57.7^2}{3} + \frac{46.6^2}{3} + \cdots + \frac{9.5^2}{3} - \text{CM} = 2303.2 - \text{CM} = 605.6978$$

$$\text{SSE} = \text{SS(Total)} - \text{SST} - \text{SSB} = 13.0489$$

$$\text{MST} = \frac{\text{SST}}{p-1} = \frac{.1911}{2} = .10 \qquad \text{MSB} = \frac{\text{SSB}}{b-1} = \frac{605.6978}{5} = 121.14$$

$$\text{MSE} = \frac{\text{SSE}}{n-p-b+1} = \frac{13.0489}{10} = 1.30$$

$$F_T = \frac{\text{MST}}{\text{MSE}} = \frac{.10}{1.30} = .08 \qquad F_B = \frac{\text{MSB}}{\text{MSE}} = \frac{121.14}{1.30} = 93.18$$

c. H_0: $\mu_1 = \mu_2 = \mu_3$
H_a: At least two of the three means differ

Test statistic: $F = 0.08$

Rejection Region: For $\alpha = .05$, $\nu_1 = p - 1 = 2$, $\nu_2 = n - p - b + 1 = 10$, $F_{.05} = 4.10$
Reject H_0 if $F > 4.10$

Conclusion: Do not reject H_0 at $\alpha = .05$. There is insufficient evidence to indicate a difference in the mean estimates of beer brand market shares produced by the three auditing methods.

d. The form of the 95% confidence interval is:

$$(\bar{T}_1 - \bar{T}_2) \pm t_{\alpha/2} s\sqrt{\frac{2}{b}} \quad \text{where } s = \sqrt{1.30} = 1.14$$

$t_{.025} = 2.228$ with df $= n - p - b + 1 = 10$

$$\bar{T}_1 = \frac{57.4}{6} = 9.567, \qquad \bar{T}_2 = \frac{58.8}{6} = 9.8$$

The 95% confidence interval is:

$$\Rightarrow (9.567 - 9.8) \pm 2.228(1.14)\sqrt{\frac{2}{6}} \Rightarrow -.233 \pm 1.466 \Rightarrow (-1.699, 1.233)$$

11.81 First, we need to calculate the treatment totals:

$$T_1 = \bar{T}_1 n_1 = 16.75(122) = 2043.5$$
$$T_2 = \bar{T}_2 n_2 = 18.35(135) = 2477.25$$
$$T_3 = \bar{T}_3 n_3 = 18.88(88) = 1661.44$$

Thus, $\sum y = 2043.5 + 2477.25 + 1661.44 = 6182.19$

$\text{CM} = (\sum y)^2/n = (6182.19)^2/345 = 110{,}781.0817$

Let groups 1, 2, and 3 represent no training, computer assisted training, and computer training plus workshop, respectively. Then,

$$\text{SST} = \frac{T_1^2}{n_1} + \frac{T_2^2}{n_2} + \frac{T_3^2}{n_3} - \text{CM} = \frac{(2043.5)^2}{122} + \frac{(2477.5)^2}{135} + \frac{(1661.44)^2}{88} - \text{CM}$$

$$= 111{,}054.1497 - \text{CM} = 273.068$$

$$\text{SSE} = (n_1 - 1)s_1^2 + (n_2 - 1)s_2^2 + (n_3 - 1)s_3^2$$
$$= (121)(1.37)^2 + (134)(1.33)^2 + (87)(1.20)^2 = 589.4175$$

$\text{SS(Total)} = \text{SST} + \text{SSE} = 273.068 + 589.4175 = 862.4855$

To complete the analysis, we calculate:

$\text{df(Treatments)} = p - 1 = 3 - 1 = 2$
$\text{df(Error)} = n - p = 345 - 3 = 342$
$\text{df(Total)} = n - 1 = 345 - 1 = 344$
$\text{MST} = \text{SST}/(p - 1) = 273.068/2 = 136.534$
$\text{MSE} = \text{SSE}/(n - p) = 589.4175/342 = 1.7234$
$F = \text{MST}/\text{MSE} = 79.22$

The complete ANOVA table is shown below:

Source	df	SS	MS	F
Treatments	2	273.068	136.534	79.22
Error	342	589.4175	1.7234	
Total	344	862.4855		

H_0: $\mu_1 = \mu_2 = \mu_3$
H_a: At least two of the means differ

Test statistic: $F = 79.22$

Rejection region: $\alpha = .05$, $\nu_1 = 2$, $\nu_2 = 342$, $F_{.05} = 3.00$
Reject H_0 if $F > 3.00$

Conclusion: Reject H_0 at $\alpha = .05$. There is sufficient evidence to indicate a difference among the mean effectiveness of performance appraisal training of the three groups.

Because there is evidence of differences among the treatment means, we will run a multiple comparison using Scheffe's method. We have MSE $= 1.7234$, $p = 3$, and $\nu =$ df(Error) $= 342$. Also, $n_1 = 122$, $n_2 = 135$, and $n_3 = 88$. There are 3 different critical values:

For $\alpha = .05$, $F_{.05}(2, 342) \approx 3.00$ and

$$S_{12} = \sqrt{(p-1)F_{.05}(2,\ 342)\ \text{MSE}\left(\frac{1}{n_1} + \frac{1}{n_2}\right)} = \sqrt{(2)(3.00)(1.7234)\left(\frac{1}{122} + \frac{1}{135}\right)} = .402$$

$$S_{13} = \sqrt{(2)(3.00)(1.7234)\left(\frac{1}{122} + \frac{1}{88}\right)} = .450$$

$$S_{23} = \sqrt{(2)(3.00)(1.7234)\left(\frac{1}{135} + \frac{1}{88}\right)} = .441$$

The sample means arranged in order are:

NT	CAT	CT+
16.75	18.35	18.88

There are significant differences between all pairs of means.

11.83 a. The complete second-order model is:

$$E(y) = \beta_0 + \beta_1 x_1 + \beta_2 x_2 + \beta_3 x_1 x_2 + \beta_4 x_1^2 + \beta_5 x_2^2$$

b. The prediction equation is:

$$\hat{y} = 29.85625 + .56x_1 - .1625\,x_2 - .1135\,x_1 x_2 - .275\,x_1^2 - .23125x_2^2$$

c. The SSE's are different because the models are different. In Exercise 11.82, a model of higher than second order was fit.

d. $R^2 = .8421$. This implies that 84.21% of the sample variation in the productivity scores is explained by the complete second order model containing arrival rate and temperature.

e. H_0: $\beta_6 = \beta_7 = \cdots = \beta_{15} = 0$
H_a: At least one of the betas differs from 0

Test statistic: $F = \dfrac{(\text{SSE}_R - \text{SSE}_C)/10}{\text{SSE}_C/[32 - (15 + 1)]} = \dfrac{(24.5332 - 5.4)/10}{5.4/16} = 5.67$

Rejection region: $\alpha = .05$, $\nu_1 = 10$, $\nu_2 = 16$, $F_{.05} = 2.49$
Reject H_0 if $F > 2.49$

Conclusion: Reject H_0 at $\alpha = .05$. There is sufficient evidence to indicate the complete model provides more information for predicting y than the second-order model.

11.85 a. Recall that $\bar{y} = \frac{\sum y}{n} \Rightarrow \sum y = n\bar{y}$; therefore,

BS Interaction Totals

$B_1S_1 = 20(160.3) = 3206$
$B_1S_2 = 20(155.5) = 3110$
$B_2S_1 = 20(173.05) = 3461$
$B_2S_2 = 20(149.50) = 2990$

Buyer Condition (B) Totals

$B_1 = B_1S_1 + B_1S_2 = 3206 + 3110 = 6316$
$B_2 = B_2S_1 + B_2S_2 = 3461 + 2990 = 6451$

Seller Condition (S) Totals

$S_1 = B_1S_1 + B_2S_1 = 3206 + 3461 = 6667$
$S_2 = B_1S_2 + B_2S_2 = 3110 + 2990 = 6100$

$$\sum y = B_1S_1 + B_1S_2 + B_2S_1 + B_2S_2 = 3206 + 3110 + 3461 + 2990 = 12{,}767$$

$$\text{CM} = \frac{(\sum y)^2}{n} = \frac{(12{,}767)^2}{80} = 2{,}037{,}453.6125$$

$$\text{SS}(B) = \frac{(6316)^2 + (6451)^2}{2(20)} - \text{CM} = 2{,}037{,}681.425 - \text{CM} = 227.8125$$

$$\text{SS}(S) = \frac{(6667)^2 + (6100)^2}{2(20)} - \text{CM} = 2{,}041{,}472.225 - \text{CM} = 4018.6125$$

$$\text{SS}(BS) = \frac{(3206)^2 + (3110)^2 + (3461)^2 + (2990)^2}{20} - \text{SS}(B) - \text{SS}(S) - \text{CM}$$
$$= 2{,}043{,}457.85 - 227.8125 - 4018.6125 - 2{,}037{,}453.6125 = 1757.8125$$

Since $s = \sqrt{\text{MSE}} \approx 21.00 \Rightarrow \text{MSE} \approx 441.0$

$$\text{MSE} = \frac{\text{SSE}}{76} \Rightarrow \text{SSE} \approx 441(76) = 33{,}516.0$$
$$\text{SS(Total)} = \text{SS}(B) + \text{SS}(S) + \text{SS}(BS) + \text{SSE}$$
$$= 227.8125 + 4018.6125 + 1757.8125 + 33{,}516 = 39{,}520.2374$$

$$\text{MS}(B) = \frac{\text{SS}(B)}{1} = \frac{227.8125}{1} = 227.8125$$

$$\text{MS}(S) = \frac{\text{SS}(S)}{1} = \frac{4018.6125}{1} = 4018.6125$$

$$\text{MS}(BS) = \frac{\text{SS}(BS)}{1} = \frac{1757.8125}{1} = 1757.8125$$

The complete ANOVA table is:

Source	df	SS	MS	F
Buyers (B)	1	227.8125	227.8125	.52
Sellers (S)	1	4018.6125	4018.6125	9.11
BS	1	1757.8125	1757.8125	3.99
Error	76	33516.0000	441.0000	
Total	79	39520.2375		

b. It does appear that the mean initial offer depends on the factor-level combination of buyer condition and seller condition. The factors seem to interact because the lines are not parallel.

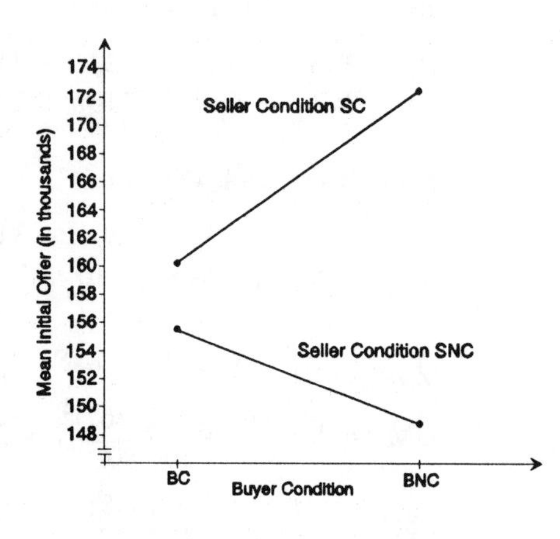

c. H_0: There is no interaction between buyer and seller
H_a: Buyer and Seller interact

Test statistic: $F = \dfrac{\text{MS}(BS)}{\text{MSE}} = 3.99$

Rejection region: $\alpha = .10,\ \nu_1 = 1,\ \nu_2 = 76,\ F_{.10} \approx 2.79$
Reject H_0 if $F > 2.79$

Conclusion: Reject H_0 at $\alpha = .10$. There is sufficient evidence to indicate that buyer condition and seller condition interact.

d. The complete and reduced models for testing for interaction in this 2×2 factorial design are:

Complete model: $E(y) = \beta_0 + \beta_1 x_1 + \beta_2 x_2 + \beta_3 x_1 x_2$

where $x_1 = \begin{cases} 1 & \text{if } BC \\ 0 & \text{if } BNC \end{cases} \quad x_2 = \begin{cases} 1 & \text{if } SC \\ 0 & \text{if } SNC \end{cases}$

Reduced model: $E(y) = \beta_0 + \beta_1 x_1 + \beta_2 x_2$

e. A 90% confidence interval for the difference between the means of cells B_1S_1 and B_2S_1 is given by:

$$(\bar{y}_{11} - \bar{y}_{21}) \pm t_{.05(76\ df)} \sqrt{MSE\left(\frac{2}{r}\right)}$$

where $\bar{y}_{11}$ = sample mean for cell B_1S_1
$\bar{y}_{21}$ = sample mean for cell B_2S_1

Substitution yields:

$$(160.3 - 173.05) \pm 1.645\sqrt{441\left(\frac{2}{20}\right)} \Rightarrow 12.75 \pm 10.924 \text{ or } (-23.674, -1.826)$$

We estimate that the mean initial offer of the BC/SC group is less than the corresponding mean for the BNC/SC group by between 1.826 and 23.674 thousand dollars. This refutes the researcher's hypothesis of no difference between the means of these two groups.

11.87 a. To test whether the model contributes information for the prediction of y, we test:

H_0: $\beta_1 = \beta_2 = \beta_3 = \beta_4 = \beta_5 = 0$
H_a: At least one β differs from 0

Test statistic: $F = 33.863$ (from printout)

Rejection region: Reject H_0 if p-value $< \alpha = .05$
p-value $= .0001$

Conclusion: Reject H_0 at $\alpha = .05$. There is sufficient evidence to indicate the model contributes information for the prediction of y, worker productivity.

b. The value $R^2 = .9338$ indicates that 93.38% of the sample variability of the y-values is explained by the model. In other words, there is a 93.38% reduction in the error of prediction by using the least squares equation, instead of $\bar{y}$, to predict y.

c. A large value of R^2 does not necessarily imply that the model is useful for prediction. However, in part (a) we performed a statistical test of the utility of the model and concluded that the second-order model is a good predictor of productivity.

d. A small value of R^2 does not necessarily imply that the model is not useful for prediction; a statistical test of model adequacy should always be performed. We could increase the value of R^2 by adding other independent variables to the model that are related to productivity, or, possibly, by changing the form of the model itself.

e. To test whether interaction exists between the two factors, we test:

H_0: $\beta_5 = 0$
H_a: $\beta_5 \neq 0$

Test statistic: $t = 0.043$ (from printout)

Rejection region: Reject H_0 if p-value $< \alpha = .05$
p-value $= .9663$

Conclusion: Do not reject H_0 at $\alpha = .05$. There is insufficient evidence to indicate the interaction exists between the two factors.

f. The parameter estimates are given in the column labelled PARAMETER ESTIMATE, and the corresponding prediction equation is:

$$\hat{y} = -4026.64 + 385.08x_1 - 24.67x_1^2 + 661.58x_2 - 37.17x_2^2 + .25x_1x_2$$

g. When $x_2 = 8$, the prediction equation is:

$$\hat{y} = -4026.64 + 385.08x_1 - 24.67x_1^2 + 661.58(8) - 37.17(8)^2 + .25x_1(8)$$

or $$\hat{y} = -1112.88 + 387.08x_1 - 24.67x_1^2$$

Similarly, when $x_2 = 9$, $\hat{y} = -1083.19 + 387.33x_1 - 24.67x_1^2$

and when $x_2 = 10$, $\hat{y} = -1127.84 + 387.58x_1 - 24.67x_1^2$

The graphs of predicted productivity $\hat{y}$ as a function of Pay rate x_1 are shown below.

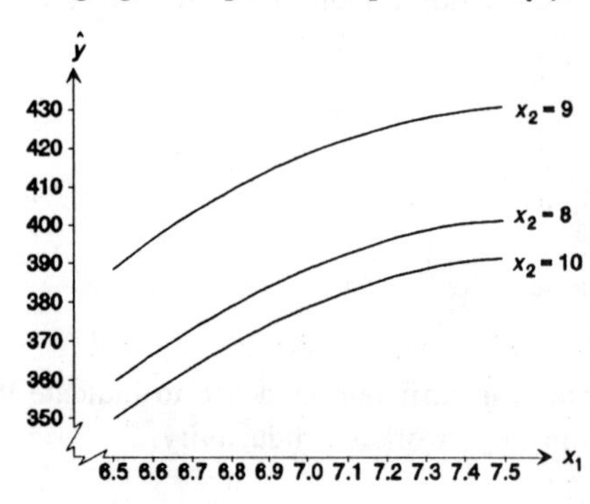

Appendix A

The Mechanics of Multiple Regression Analysis

A.1 a. Since **A** is a 2 × 2 matrix and **B** is a 2 × 2 matrix, the product **AB** will be a 2 × 2 matrix, which we will call **S**:

$$\mathbf{AB} = \mathbf{S}$$

$$\begin{bmatrix} 3 & 0 \\ -1 & 4 \end{bmatrix} \begin{bmatrix} 2 & 1 \\ 0 & -1 \end{bmatrix} = \begin{bmatrix} s_{11} & s_{12} \\ s_{21} & s_{22} \end{bmatrix}$$

To determine the elements of **S**, the following calculations are required:

$$\begin{aligned} s_{11} &= (3)(2) + (0)(0) = 6 \\ s_{12} &= (3)(1) + (0)(-1) = 3 \\ s_{21} &= (-1)(2) + (4)(0) = -2 \\ s_{22} &= (-1)(1) + (4)(-1) = -5 \end{aligned}$$

Hence, $\mathbf{AB} = \begin{bmatrix} 6 & 3 \\ -2 & -5 \end{bmatrix}$

b. $\mathbf{AC} = \begin{bmatrix} 3 & 0 \\ -1 & 4 \end{bmatrix} \begin{bmatrix} 1 & 0 & 3 \\ -2 & 1 & 2 \end{bmatrix} = \begin{bmatrix} s_{11} & s_{12} & s_{13} \\ s_{21} & s_{22} & s_{23} \end{bmatrix}$

where

$$\begin{aligned} s_{11} &= (3)(1) + (0)(-2) = 3 \\ s_{12} &= (3)(0) + (0)(1) = 0 \\ s_{13} &= (3)(3) + (0)(2) = 9 \\ s_{21} &= (-1)(1) + (4)(-2) = -9 \\ s_{22} &= (-1)(0) + (4)(1) = 4 \\ s_{23} &= (-1)(3) + (4)(2) = 5 \end{aligned}$$

Hence, $\mathbf{AC} = \begin{bmatrix} 3 & 0 & 9 \\ -9 & 4 & 5 \end{bmatrix}$

c. $\mathbf{BA} = \begin{bmatrix} 2 & 1 \\ 0 & -1 \end{bmatrix} \begin{bmatrix} 3 & 0 \\ -1 & 4 \end{bmatrix} = \begin{bmatrix} s_{11} & s_{12} \\ s_{21} & s_{22} \end{bmatrix}$

$$s_{11} = (2)(3) + (1)(-1) = 5$$
$$s_{12} = (2)(0) + (1)(4) = 4$$
$$s_{21} = (0)(3) + (-1)(-1) = 1$$
$$s_{22} = (0)(0) + (-1)(4) = -4$$

Hence, $\mathbf{BA} = \begin{bmatrix} 5 & 4 \\ 1 & -4 \end{bmatrix}$

A.3 a. If matrix **A** has dimensions $r \times d$ and matrix **B** has dimensions $d \times c$, then the product matrix **AB** has dimensions $r \times c$. In this exercise, $r = 3$, $d = 2$, and $c = 4$:

A	**B**
3×2	2×4

Thus, **AB** is a 3×4 matrix.

b. In order to multiply two matrices, the two inner dimension numbers must be equal:

B	**A**
2×4	3×2

In this exercise, the number of columns of **B** (4) does not equal the number of rows of **A** (3); thus, the product **BA** does not exist.

A.5 a. $\mathbf{AB} = \begin{bmatrix} 1 & 0 & 0 \\ 0 & 3 & 0 \\ 0 & 0 & 2 \end{bmatrix} \begin{bmatrix} 2 & 3 \\ -3 & 0 \\ 4 & -1 \end{bmatrix} = \begin{bmatrix} s_{11} & s_{12} \\ s_{21} & s_{22} \\ s_{31} & s_{32} \end{bmatrix}$

where
$$s_{11} = (1)(2) + (0)(-3) + (0)(4) = 2$$
$$s_{12} = (1)(3) + (0)(0) + (0)(-1) = 3$$
$$s_{21} = (0)(2) + (3)(-3) + (0)(4) = -9$$
$$s_{22} = (0)(3) + (3)(0) + (0)(-1) = 0$$
$$s_{31} = (0)(2) + (0)(-3) + (2)(4) = 8$$
$$s_{32} = (0)(3) + (0)(0) + (2)(-1) = -2$$

Hence, $\mathbf{AB} = \begin{bmatrix} 2 & 3 \\ -9 & 0 \\ 8 & -2 \end{bmatrix}$

b. $\mathbf{CA} = \begin{bmatrix} 3 & 0 & 2 \end{bmatrix} \begin{bmatrix} 1 & 0 & 0 \\ 0 & 3 & 0 \\ 0 & 0 & 2 \end{bmatrix} = \begin{bmatrix} s_{11} & s_{12} & s_{13} \end{bmatrix}$

where $s_{11} = (3)(1) + (0)(0) + (2)(0) = 3$
$s_{12} = (3)(0) + (0)(3) + (2)(0) = 0$
$s_{13} = (3)(0) + (0)(0) + (2)(2) = 4$

Thus, $\mathbf{CA} = \begin{bmatrix} 3 & 0 & 4 \end{bmatrix}$

c. $\mathbf{CB} = \begin{bmatrix} 3 & 0 & 2 \end{bmatrix} \begin{bmatrix} 2 & 3 \\ -3 & 0 \\ 4 & -1 \end{bmatrix} = \begin{bmatrix} s_{11} & s_{12} \end{bmatrix}$

where $s_{11} = (3)(2) + (0)(-3) + (2)(4) = 14$
$s_{12} = (3)(3) + (0)(0) + (2)(-1) = 7$

Thus, $\mathbf{CB} = \begin{bmatrix} 14 & 7 \end{bmatrix}$

A.7 a. To obtain the product **IA**, the number of rows of **A** (2) must equal the number of columns of **I**:

$$\begin{matrix} \mathbf{I} & & \mathbf{A} \\ 2 \times 2 & & 2 \times 3 \end{matrix}$$

An identity matrix is always square, so I will have 2 rows and 2 columns:

$$\mathbf{I} = \begin{bmatrix} 1 & 0 \\ 0 & 1 \end{bmatrix}$$

b. $\mathbf{IA} = \begin{bmatrix} 1 & 0 \\ 0 & 1 \end{bmatrix} \begin{bmatrix} 3 & 0 & 2 \\ -1 & 1 & 4 \end{bmatrix} = \begin{bmatrix} 3 & 0 & 2 \\ -1 & 1 & 4 \end{bmatrix} = \mathbf{A}$

c. To find the product **AI**, the number of rows and columns of **I** must equal the number of columns of **A**(3):

$$\begin{matrix} \mathbf{A} & & \mathbf{I} \\ 2 \times 3 & & 3 \times 3 \end{matrix}$$

Hence, **I** is the 3×3 identity matrix.

$$\mathbf{I} = \begin{bmatrix} 1 & 0 & 0 \\ 0 & 1 & 0 \\ 0 & 0 & 1 \end{bmatrix}$$

d. $\mathbf{AI} = \begin{bmatrix} 3 & 0 & 2 \\ -1 & 1 & 4 \end{bmatrix} \begin{bmatrix} 1 & 0 & 0 \\ 0 & 1 & 0 \\ 0 & 0 & 1 \end{bmatrix} = \begin{bmatrix} 3 & 0 & 2 \\ -1 & 1 & 4 \end{bmatrix} = \mathbf{A}$

A.9 It is necessary to show that $\mathbf{AA}^{-1} = \mathbf{I}$ and $\mathbf{A}^{-1}\mathbf{A} = \mathbf{I}$:

$$\mathbf{AA}^{-1} = \begin{bmatrix} 12 & 0 & 0 & 8 \\ 0 & 12 & 0 & 0 \\ 0 & 0 & 8 & 0 \\ 8 & 0 & 0 & 8 \end{bmatrix} \begin{bmatrix} 1/4 & 0 & 0 & -1/4 \\ 0 & 1/12 & 0 & 0 \\ 0 & 0 & 1/8 & 0 \\ -1/4 & 0 & 0 & 3/8 \end{bmatrix} = \begin{bmatrix} 1 & 0 & 0 & 0 \\ 0 & 1 & 0 & 0 \\ 0 & 0 & 1 & 0 \\ 0 & 0 & 0 & 1 \end{bmatrix} = \mathbf{I}$$

where $\mathbf{I}$ is the 4×4 identity matrix. Similarly, it is easy to verify that $\mathbf{A}^{-1}\mathbf{A} = \mathbf{I}$.

A.11 It is necessary to show that $\mathbf{DD}^{-1} = \mathbf{I}$ and $\mathbf{D}^{-1}\mathbf{D} = \mathbf{I}$:

$$\mathbf{DD}^{-1} = \begin{bmatrix} d_{11} & 0 & 0 & \cdots & 0 \\ 0 & d_{22} & 0 & \cdots & 0 \\ 0 & 0 & d_{33} & \cdots & 0 \\ \vdots & \vdots & \vdots & \vdots\vdots\vdots & \vdots \\ 0 & 0 & 0 & \cdots & d_{nn} \end{bmatrix} \begin{bmatrix} 1/d_{11} & 0 & 0 & \cdots & 0 \\ 0 & 1/d_{22} & 0 & \cdots & 0 \\ 0 & 0 & 1/d_{33} & \cdots & 0 \\ \vdots & \vdots & \vdots & \vdots\vdots\vdots & \vdots \\ 0 & 0 & 0 & \cdots & 1/d_{nn} \end{bmatrix} = \begin{bmatrix} 1 & 0 & 0 & \cdots & 0 \\ 0 & 1 & 0 & \cdots & 0 \\ 0 & 0 & 1 & \cdots & 0 \\ \vdots & \vdots & \vdots & \vdots\vdots\vdots & \vdots \\ 0 & 0 & 0 & \cdots & 1 \end{bmatrix} = \mathbf{I}$$

where $\mathbf{I}$ is the $n \times n$ identity matrix. Upon multiplication, we find that $\mathbf{DD}^{-1} = \mathbf{I}$ when all $d_{ii} \neq 0$. Similarly, it is easy to verify that $\mathbf{D}^{-1}\mathbf{D} = \mathbf{I}$ when all $d_{ii} \neq 0$.

A.13 The inverse of a 2×2 matrix

$$\mathbf{A} = \begin{bmatrix} a & b \\ c & d \end{bmatrix} \text{ is } \mathbf{A}^{-1} = \begin{bmatrix} \frac{d}{ad - bc} & \frac{-b}{ad - bc} \\ \frac{-c}{ad - bc} & \frac{a}{ad - bc} \end{bmatrix}$$

Thus, for $\mathbf{A} = \begin{bmatrix} 2 & -1 \\ 2 & 3 \end{bmatrix}$

$$\mathbf{A}^{-1} = \begin{bmatrix} \frac{3}{2(3) - (-1)(2)} & \frac{-(-1)}{2(3) - (-1)(2)} \\ \frac{-2}{2(3) - (-1)(2)} & \frac{2}{2(3) - (-1)(2)} \end{bmatrix} = \begin{bmatrix} 3/8 & 1/8 \\ -1/4 & 1/4 \end{bmatrix}$$

A.15 We will first rewrite the system of equations in standard form:

$$10v_1 + 0v_2 + 20v_3 = 60$$
$$0v_1 + 20v_2 + 0v_3 = 60$$
$$20v_1 + 0v_2 + 68v_3 = 176$$

a. A is the matrix of coefficients of the v's:

$$\mathbf{A} = \begin{bmatrix} 10 & 0 & 20 \\ 0 & 20 & 0 \\ 20 & 0 & 68 \end{bmatrix} \qquad \mathbf{V} = \begin{bmatrix} v_1 \\ v_2 \\ v_3 \end{bmatrix} \qquad \mathbf{G} = \begin{bmatrix} 60 \\ 60 \\ 176 \end{bmatrix}$$

b.

$$\mathbf{AA}^{-1} = \begin{bmatrix} 10 & 0 & 20 \\ 0 & 20 & 0 \\ 20 & 0 & 68 \end{bmatrix} \begin{bmatrix} 17/70 & 0 & -1/14 \\ 0 & 1/20 & 0 \\ -1/14 & 0 & 1/28 \end{bmatrix}$$

$$= \begin{bmatrix} 17/7 + 0 - 10/7 & 0 + 0 + 0 & -5/7 + 0 + 5/7 \\ 0 + 0 + 0 & 0 + 1 + 0 & 0 + 0 + 0 \\ 34/7 + 0 - 34/7 & 0 + 0 + 0 & -10/7 + 0 + 17/7 \end{bmatrix}$$

$$= \begin{bmatrix} 1 & 0 & 0 \\ 0 & 1 & 0 \\ 0 & 0 & 1 \end{bmatrix}$$

c.

$$\mathbf{V} = \mathbf{A}^{-1}\mathbf{G} = \begin{bmatrix} 17/70 & 0 & -1/14 \\ 0 & 1/20 & 0 \\ -1/14 & 0 & 1/28 \end{bmatrix} \begin{bmatrix} 60 \\ 60 \\ 176 \end{bmatrix} = \begin{bmatrix} 2 \\ 3 \\ 2 \end{bmatrix}$$

The solution may be verified by substitution of $v_1 = 2$, $v_2 = 3$, and $v_3 = 2$ into the original system of equations:

$$10(0) + 20(2) - 60 = 0$$
$$20(3) - 60 = 0$$
$$20(0) + 68(2) - 176 = 0$$

A.17 a. $\mathbf{Y}$ is a 6×1 column matrix and $\mathbf{X}$ is a 6×2 data matrix:

$$\mathbf{Y} = \begin{bmatrix} 1 \\ 2 \\ 2 \\ 3 \\ 5 \\ 5 \end{bmatrix} \qquad \mathbf{X} = \begin{bmatrix} 1 & 1 \\ 1 & 2 \\ 1 & 3 \\ 1 & 4 \\ 1 & 5 \\ 1 & 6 \end{bmatrix}$$

b. $$\mathbf{X}'\mathbf{X} = \begin{bmatrix} 1 & 1 & 1 & 1 & 1 & 1 \\ 1 & 2 & 3 & 4 & 5 & 6 \end{bmatrix} \begin{bmatrix} 1 & 1 \\ 1 & 2 \\ 1 & 3 \\ 1 & 4 \\ 1 & 5 \\ 1 & 6 \end{bmatrix} = \begin{bmatrix} 6 & 21 \\ 21 & 91 \end{bmatrix}$$

$$\mathbf{X}'\mathbf{Y} = \begin{bmatrix} 1 & 1 & 1 & 1 & 1 & 1 \\ 1 & 2 & 3 & 4 & 5 & 6 \end{bmatrix} \begin{bmatrix} 1 \\ 2 \\ 2 \\ 3 \\ 5 \\ 5 \end{bmatrix} = \begin{bmatrix} 18 \\ 78 \end{bmatrix}$$

c. $$(\mathbf{X}'\mathbf{X})^{-1}(\mathbf{X}'\mathbf{X}) = \begin{bmatrix} 13/15 & -7/35 \\ -7/35 & 2/35 \end{bmatrix} \begin{bmatrix} 6 & 21 \\ 21 & 91 \end{bmatrix} = \begin{bmatrix} 1 & 0 \\ 0 & 1 \end{bmatrix} = \mathbf{I};$$

similarly, $(\mathbf{X}'\mathbf{X})(\mathbf{X}'\mathbf{X})^{-1} = \mathbf{I}$

d. To obtain the least squares estimates, we computer $\hat{\beta} = (\mathbf{X}'\mathbf{X})^{-1}\mathbf{X}'\mathbf{Y}$:

$$\hat{\beta} = \begin{bmatrix} 13/15 & -7/35 \\ -7/35 & 2/35 \end{bmatrix} \begin{bmatrix} 18 \\ 78 \end{bmatrix} = \begin{bmatrix} 0 \\ .8571 \end{bmatrix}$$

Thus, $\hat{\beta}_0 = 0$ and $\hat{\beta}_1 = .8571$.

e. The prediction equation is:

$$\hat{y} = \hat{\beta}_0 + \hat{\beta}_1 x = 0 + .8571x = .8571x$$

The identical solution was obtained in Exercise 3.6.

A.19 From Exercise A.16,

$$\mathbf{X} = \begin{bmatrix} 1 & -2 \\ 1 & -1 \\ 1 & 0 \\ 1 & 1 \\ 1 & 2 \end{bmatrix} \qquad \mathbf{Y} = \begin{bmatrix} 4 \\ 3 \\ 3 \\ 1 \\ -1 \end{bmatrix} \qquad \mathbf{X'X} = \begin{bmatrix} 5 & 0 \\ 0 & 10 \end{bmatrix}$$

$$\mathbf{X'X}^{-1} = \begin{bmatrix} 1/5 & 0 \\ 0 & 1/10 \end{bmatrix} \qquad \mathbf{X'Y} = \begin{bmatrix} 10 \\ -12 \end{bmatrix} \qquad \hat{\beta} = \begin{bmatrix} 2 \\ -1.2 \end{bmatrix}$$

We want to test:

$$H_0\!: \beta_1 = 0$$
$$H_a\!: \beta_1 \neq 0$$

Test statistic: $t = \dfrac{\hat{\beta}_1}{s\sqrt{c_{11}}}$

where $s = \sqrt{\dfrac{\text{SSE}}{n - (k + 1)}}$, $\text{SSE} = \mathbf{Y'Y} - \hat{\beta}'\mathbf{X'Y}$, and c_{11} is obtained from the $(\mathbf{X'X})^{-1}$ matrix.

We now compute $\mathbf{Y'Y}$: $\begin{bmatrix} 4 & 3 & 3 & 1 & -1 \end{bmatrix} \begin{bmatrix} 4 \\ 3 \\ 3 \\ 1 \\ -1 \end{bmatrix} = 36;$

then, $\hat{\beta}'\mathbf{X'Y} = \begin{bmatrix} 2 & -1.2 \end{bmatrix} \begin{bmatrix} 10 \\ -12 \end{bmatrix} = 34.4$

Hence, $\text{SSE} = \mathbf{Y'Y} - \hat{\beta}'\mathbf{X'Y} = 36 - 34.4 = 1.6$

and $s = \sqrt{\dfrac{\text{SSE}}{n - (k + 1)}} = \sqrt{\dfrac{1.6}{5 - (1 + 1)}} = \sqrt{.5333} = .7303$

If $\mathbf{X'X}^{-1} = \begin{bmatrix} 1/5 & 0 \\ 0 & 1/10 \end{bmatrix}$, then $c_{11} = 1/10$.

The test statistic is computed as follows:

$$t = \frac{-1.2}{.7303\sqrt{1/10}} = -5.196$$

Rejection region: $\alpha = .05$, df $= 3$, $t_{.025} = 3.182$;
Reject H_0 if $t < -3.182$ or $t > 3.182$

Conclusion: Reject H_0 at $\alpha = .05$. There is sufficient evidence that x contributes information for the prediction of y.

A.21 From Exercise A.18,

$$\mathbf{X} = \begin{bmatrix} 1 & -2 & 4 \\ 1 & -2 & 4 \\ 1 & -1 & 1 \\ 1 & -1 & 1 \\ 1 & 0 & 0 \\ 1 & 0 & 0 \\ 1 & 1 & 1 \\ 1 & 1 & 1 \\ 1 & 2 & 4 \\ 1 & 2 & 4 \end{bmatrix} \quad \mathbf{Y} = \begin{bmatrix} 1.1 \\ 1.3 \\ 2.0 \\ 2.1 \\ 2.7 \\ 2.8 \\ 3.4 \\ 3.6 \\ 4.1 \\ 4.0 \end{bmatrix} \quad \mathbf{X'X} = \begin{bmatrix} 10 & 0 & 20 \\ 0 & 20 & 0 \\ 20 & 0 & 68 \end{bmatrix}$$

$$\mathbf{X'X}^{-1} = \begin{bmatrix} 17/70 & 0 & -1/14 \\ 0 & 1/20 & 0 \\ -1/14 & 0 & 1/28 \end{bmatrix} \quad \mathbf{X'Y} = \begin{bmatrix} 27.1 \\ 14.3 \\ 53.1 \end{bmatrix} \quad \hat{\beta} = \begin{bmatrix} 2.7885714 \\ 0.715 \\ -0.03928571 \end{bmatrix}$$

H_0: $\beta_2 = 0$
H_a: $\beta_2 \neq 0$

Test statistic: $t = \dfrac{\hat{\beta}_2}{s\sqrt{c_{22}}}$

where $s = \sqrt{\dfrac{\text{SSE}}{n - (k + 1)}}$, SSE $= \mathbf{Y'Y} - \hat{\beta}'\mathbf{X'Y}$, and c_{22} is obtained from the $(\mathbf{X'X})^{-1}$ matrix.

We now compute $\mathbf{Y'Y}$:

$$\mathbf{Y'Y} = \begin{bmatrix} 1.1 & 1.3 & 2.0 & 2.1 & 2.7 & 2.8 & 3.4 & 3.6 & 4.1 & 4.0 \end{bmatrix} \begin{bmatrix} 1.1 \\ 1.3 \\ 2.0 \\ 2.1 \\ 2.7 \\ 2.8 \\ 3.4 \\ 3.6 \\ 4.1 \\ 4.0 \end{bmatrix}$$

$\mathbf{Y'Y} = 83.77$; then

$$\hat{\beta}'\mathbf{X'Y} = \begin{bmatrix} 2.7885714 & .715 & -.03928571 \end{bmatrix} \begin{bmatrix} 27.1 \\ 14.3 \\ 53.1 \end{bmatrix} = 83.708714$$

Hence, $\text{SSE} = \mathbf{Y'Y} - \hat{\beta}'\mathbf{X'Y} = 83.77 - 83.708714 = .061286$

and $s = \sqrt{\dfrac{\text{SSE}}{n - (k + 1)}} = \sqrt{\dfrac{.061286}{10 - (2 + 1)}} = \sqrt{.0087551} = .093569$

The test statistic is computed as follows:

$$t = \frac{-.0392857}{.093569\sqrt{1/28}} = -2.22$$

Rejection region: $\alpha = .10$, df $= 7$, $t_{.05} = 1.895$;
Reject H_0 if $t < -1.895$ or $t > 1.895$

Conclusion: Reject H_0 at $\alpha = .10$. There is sufficient evidence to indicate curvature in the model for $E(y)$.

A.23 The 90% prediction interval is given by $\hat{y} \pm t_{.05} s\sqrt{1 + a'(\mathbf{X'X})^{-1}a}$

where $\hat{y} = 2 - 1.2(1) = 0.8$,
$t_{.05} = 2.353$ (3 df),
$s = .7303$ (from Exercises A.16 and A.19)

$$a = \begin{bmatrix} 1 \\ 1 \end{bmatrix}$$

and $\mathbf{X'X} = \begin{bmatrix} 5 & 0 \\ 0 & 10 \end{bmatrix}$ (from Exercises A.16 and A.19)

Now, $a'(\mathbf{X'X})^{-1}a = \begin{bmatrix} 1 & 1 \end{bmatrix} \begin{bmatrix} 1/5 & 0 \\ 0 & 1/10 \end{bmatrix} \begin{bmatrix} 1 \\ 1 \end{bmatrix} = 0.3$

and substitution yields the desired interval:

$$0.8 \pm 2.353(.7303\sqrt{1 + 0.3} \Rightarrow 0.8 \pm 1.959 \Rightarrow (-1.159, 2.759)$$

For a future trial of the experiment with $x = 1$, we predict that the y-value will lie between -1.159 and 2.759 with 90% confidence.

A.25 The 90% prediction interval is given by $\hat{y} \pm t_{.05} s\sqrt{1 + a'(\mathbf{X'X})^{-1}a}$

where $\hat{y} = .8571(2) = 1.7142$,
$t_{.05} = 2.132$ (4 df)

$$s = \sqrt{\frac{\text{SSE}}{n - (k+1)}} = \sqrt{\frac{(\mathbf{Y'Y} - \hat{\beta}'\mathbf{X'Y})}{n - (k+1)}} = \sqrt{\frac{68 - 66.8538}{4}} = \sqrt{\frac{1.1462}{4}}$$

$$= .535304$$

$$a = \begin{bmatrix} 1 \\ 2 \end{bmatrix}$$

and $(\mathbf{X'X})$ is obtained from Exercise A.17.

$$\text{Now, } a'(\mathbf{X'X})^{-1}a = \begin{bmatrix} 1 & 2 \end{bmatrix} \begin{bmatrix} 13/15 & -7/35 \\ -7/35 & 2/35 \end{bmatrix} \begin{bmatrix} 1 \\ 2 \end{bmatrix} = .295238$$

and substitution yields the desired interval:

$$1.7142 \pm 2.132(.535304)\sqrt{1 + .295238} \Rightarrow 1.7142 \pm 1.29886 \Rightarrow (.41534, 3.01306)$$

For a future trial of the experiment with $x = 2$, we predict that the y-value will lie between .42 and 3.01, with 90% confidence.

A.27 The 90% prediction interval is given by $\hat{y} \pm t_{.05}s\sqrt{1 + a'(\mathbf{X'X})^{-1}a}$

where $\hat{y} = 2.789 + .715(1) - .039(1)^2 = 3.465$,
$t_{.05} = 1.895$ (7 df)
$s = .093569$ (from Exercises A.18 and A.21)

$$\text{and } a'(\mathbf{X'X})^{-1}a = \begin{bmatrix} 1 & 1 & 1 \end{bmatrix} \begin{bmatrix} 17/70 & 0 & -1/14 \\ 0 & 1/20 & 0 \\ -1/14 & 0 & 1/28 \end{bmatrix} \begin{bmatrix} 1 \\ 1 \\ 1 \end{bmatrix} = .1857143$$

Thus, the required interval is:

$$3.465 \pm 1.895(.093569)\sqrt{1 + .1857143} \Rightarrow 3.465 \pm .1931 \Rightarrow (3.2719, 3.6581)$$

For a future trial of the experiment with $x = 1$, we predict that the y-value will lie between 3.2719 and 3.6581, with 90% confidence.

A.29 a. $$\mathbf{Y} = \begin{bmatrix} 1.1 \\ 1.9 \\ 3.0 \\ 3.8 \\ 5.1 \\ 6.0 \end{bmatrix}, \quad \mathbf{X} = \begin{bmatrix} 1 & -5 \\ 1 & -3 \\ 1 & -1 \\ 1 & 1 \\ 1 & 3 \\ 1 & 5 \end{bmatrix}$$

b. $\mathbf{X'X} = \begin{bmatrix} 1 & 1 & 1 & 1 & 1 & 1 \\ -5 & -3 & -1 & 1 & 3 & 5 \end{bmatrix} \begin{bmatrix} 1 & -5 \\ 1 & -3 \\ 1 & -1 \\ 1 & 1 \\ 1 & 3 \\ 1 & 5 \end{bmatrix} = \begin{bmatrix} 6 & 0 \\ 0 & 70 \end{bmatrix}$

$$\mathbf{X'Y} = \begin{bmatrix} 1 & 1 & 1 & 1 & 1 & 1 \\ -5 & -3 & -1 & 1 & 3 & 5 \end{bmatrix} \begin{bmatrix} 1.1 \\ 1.9 \\ 3.0 \\ 3.8 \\ 5.1 \\ 6.0 \end{bmatrix} = \begin{bmatrix} 20.9 \\ 34.9 \end{bmatrix}$$

c. To obtain the least squares estimates, we first need to obtain $(\mathbf{X'X})^{-1}$.

$$(\mathbf{X'X})^{-1} = \begin{bmatrix} 6 & 0 \\ 0 & 70 \end{bmatrix}^{-1} = \begin{bmatrix} 1/6 & 0 \\ 0 & 1/70 \end{bmatrix}, \text{ by Theorem A.1.}$$

Then, $\hat{\beta} = (\mathbf{X'X})^{-1}\mathbf{X'Y} = \begin{bmatrix} 1/6 & 0 \\ 0 & 1/70 \end{bmatrix} \begin{bmatrix} 20.9 \\ 34.9 \end{bmatrix} = \begin{bmatrix} 3.4833333 \\ .4985714 \end{bmatrix}$

Thus, $\hat{\beta}_0 = 3.48333$ and $\hat{\beta}_1 = .49857$.

d. The prediction equation is $\hat{y} = \hat{\beta}_0 + \hat{\beta}_1 x = 3.4833 + .4986x$.

e. $\text{SSE} = \mathbf{Y'Y} - \hat{\beta}_1'(\mathbf{X'Y})$, where

$$\mathbf{Y'Y} = \begin{bmatrix} 1.1 & 1.9 & 3.0 & 3.8 & 5.1 & 6.0 \end{bmatrix} \begin{bmatrix} 1.1 \\ 1.9 \\ 3.0 \\ 3.8 \\ 5.1 \\ 6.0 \end{bmatrix} = 90.27$$

and $\hat{\beta}'(\mathbf{X'Y}) = \begin{bmatrix} 3.4833333 & .4985714 \end{bmatrix} \begin{bmatrix} 20.9 \\ 34.9 \end{bmatrix} = 90.201808$

Thus, $\text{SSE} = 90.27 - 90.201808 = .0681922$, and

$$s^2 = \text{SSE}/[n - (k + 1)] = .681922/[6 - (1 + 1)] = .0681922/4 = .0170481$$

f. $H_0: \beta_1 = 0$
$H_a: \beta_1 \neq 0$

Test statistic: $t = \dfrac{\hat{\beta}_1}{s\sqrt{c_{11}}} = \dfrac{.4986}{\sqrt{.0170481}\sqrt{1/70}} = 31.95$

[Note: The value of c_{11} is obtained from the $(\mathbf{X}'\mathbf{X})^{-1}$ matrix.]

Rejection region: $\alpha = .05$, df = 4, $t_{.025} = 2.776$
Reject H_0 if $t < -2.776$ or $t > 2.776$

Conclusion: Reject H_0 at $\alpha = .05$. There is sufficient evidence to conclude that the model is useful for predicting y.

g. $r^2 = 1 - (\text{SSE}/\text{SS}_{yy})$, where SSE = .0681922 and

$$\text{SS}_{yy} = \sum y_i^2 - \frac{\left(\sum y_i\right)^2}{n} = 90.27 - \frac{(20.9)^2}{6} = 17.4683333$$

Thus, $r^2 = 1 - (.0681922/17.4683333) = .99609$.

Over 99% of the variability of the y-values about their sample mean is accounted for by the least squares model.

h. The 90% confidence interval for $E(y)$ when $x = .5$ is given by $\ell \pm t_{.05} s\sqrt{a'(\mathbf{X}'\mathbf{X})^{-1}a}$

where $\ell = \hat{y} = 3.4833 + .4986(.5) = 3.7326$
$t_{.05} = 2.132$ (based on 4 df)
$s = \sqrt{.0170481}$ [from part (e)]

and $a'(\mathbf{X}'\mathbf{X})^{-1}a = \begin{bmatrix} 1 & .5 \end{bmatrix} \begin{bmatrix} 1/6 & 0 \\ 0 & 1/70 \end{bmatrix} \begin{bmatrix} 1 \\ .5 \end{bmatrix} = \begin{bmatrix} 1/6 & .0071429 \end{bmatrix} \begin{bmatrix} 1 \\ .5 \end{bmatrix} = .1702381$

Thus, the desired confidence interval is:

$$3.7326 \pm 2.132\sqrt{.0170481}\sqrt{.1702381} \Rightarrow 3.7326 \pm .11486 \Rightarrow (3.6177, 3.8475)$$

The mean value of y when $x = .5$ is estimated to lie within the interval (3.6177, 3.8475), with 90% confidence.

A.31 a. $\mathbf{Y} = \begin{bmatrix} 1.1 \\ .5 \\ 1.8 \\ 2.0 \\ 2.0 \\ 2.9 \\ 3.8 \\ 3.4 \\ 4.1 \\ 5.0 \\ 5.0 \\ 5.8 \end{bmatrix}$, $\mathbf{X} = \begin{bmatrix} 1 & 1 \\ 1 & 1 \\ 1 & 2 \\ 1 & 2 \\ 1 & 3 \\ 1 & 3 \\ 1 & 4 \\ 1 & 4 \\ 1 & 5 \\ 1 & 5 \\ 1 & 6 \\ 1 & 6 \end{bmatrix}$

b. $$\mathbf{X'X} = \begin{bmatrix} 1 & 1 & 1 & 1 & 1 & 1 & 1 & 1 & 1 & 1 & 1 & 1 \\ 1 & 1 & 2 & 2 & 3 & 3 & 4 & 4 & 5 & 5 & 6 & 6 \end{bmatrix} \begin{bmatrix} 1 & 1 \\ 1 & 1 \\ 1 & 2 \\ 1 & 2 \\ 1 & 3 \\ 1 & 3 \\ 1 & 4 \\ 1 & 4 \\ 1 & 5 \\ 1 & 5 \\ 1 & 6 \\ 1 & 6 \end{bmatrix} = \begin{bmatrix} 12 & 42 \\ 42 & 182 \end{bmatrix}$$

$$\mathbf{X'Y} = \begin{bmatrix} 1 & 1 & 1 & 1 & 1 & 1 & 1 & 1 & 1 & 1 & 1 & 1 \\ 1 & 1 & 2 & 2 & 3 & 3 & 4 & 4 & 5 & 5 & 6 & 6 \end{bmatrix} \begin{bmatrix} 1.1 \\ .5 \\ 1.8 \\ 2.0 \\ 2.0 \\ 2.9 \\ 3.8 \\ 3.4 \\ 4.1 \\ 5.0 \\ 5.0 \\ 5.8 \end{bmatrix} = \begin{bmatrix} 37.4 \\ 163 \end{bmatrix}$$

c. The elements of the **X′X** matrix for two replications are equal to the elements of the **X′X** matrix for a single replication, multiplied by a factor of 2:

$$\mathbf{X'X} = \begin{bmatrix} 6 & 21 \\ 21 & 91 \end{bmatrix} \text{ for single replication;}$$

$$\mathbf{X'X} = \begin{bmatrix} 12 & 42 \\ 42 & 182 \end{bmatrix} \text{ for two replications.}$$

d. $$(\mathbf{X'X})^{-1} = \begin{bmatrix} 13/15 & -7/35 \\ -7/35 & 2/35 \end{bmatrix} \text{ for one replication;}$$

$$(\mathbf{X'X})^{-1} = \begin{bmatrix} 13/30 & -7/70 \\ -7/70 & 2/70 \end{bmatrix} \text{ for two replications.}$$

Verification: Show that

$$(\mathbf{X'X})^{-1}(\mathbf{X'X}) = \mathbf{I} \text{ and } (\mathbf{X'X})(\mathbf{X'X})^{-1} = \mathbf{I}$$

for two replications:

$$(\mathbf{X'X})^{-1}(\mathbf{X'X}) = \begin{bmatrix} 13/30 & -7/70 \\ -7/70 & 2/70 \end{bmatrix} \begin{bmatrix} 12 & 42 \\ 42 & 182 \end{bmatrix} = \begin{bmatrix} 1 & 0 \\ 0 & 1 \end{bmatrix} = \mathbf{I}$$

Similarly, $$(\mathbf{X'X})(\mathbf{X'X})^{-1} = \begin{bmatrix} 12 & 42 \\ 42 & 182 \end{bmatrix} \begin{bmatrix} 13/30 & -7/70 \\ -7/70 & 2/70 \end{bmatrix} = \begin{bmatrix} 1 & 0 \\ 0 & 1 \end{bmatrix} = \mathbf{I}$$

e. We first obtain the least squares estimates:

$$\hat{\beta} = (\mathbf{X'X})^{-1}\mathbf{X'Y} = \begin{bmatrix} 13/30 & -7/70 \\ -7/70 & 2/70 \end{bmatrix} \begin{bmatrix} 37.4 \\ 163 \end{bmatrix} = \begin{bmatrix} -.0933333 \\ .9171429 \end{bmatrix}$$

Thus, the prediction equation is $\hat{y} = \hat{\beta}_0 + \hat{\beta}_1 x = -.09333 + .91714x$.

f. $\text{SSE} = \mathbf{Y'Y} - \hat{\beta}'\mathbf{X'Y} = 147.56 - 146.00363 = 1.55637;$

$s^2 = \text{SSE}/[n - (k + 1)] = 1.55637/(12 - 2) = 1.55637/10 = .155637$

g. $H_0: \beta_1 = 0$
$H_a: \beta_1 \neq 0$

Test statistic: $$t = \frac{\hat{\beta}_1}{s\sqrt{c_{11}}}$$

where $s = \sqrt{.115637}$ from part (f) and $c_{11} = 2/70$ from the $(\mathbf{X'X})^{-1}$ matrix in part (d).

Thus, $t = \dfrac{.9171429}{\sqrt{.155637}\sqrt{2/70}} = 13.75$

Rejection region: $\alpha = .05$, df $= 10$, $t_{.025} = 2.228$
Reject H_0 if $t < -2.228$ or $t > 2.228$

Conclusion: Reject H_0 at $\alpha = .05$. The data provide sufficient information to indicate that x contributes information for the prediction of y.

h. $r^2 = 1 - (\text{SSE}/\text{SS}_{yy})$, where

$\text{SSE} = 1.55637$ [from part (f)]

and

$$\text{SS}_{yy} = \sum y_i^2 - \frac{\left(\sum y_i\right)^2}{n} = 147.56 - \frac{(34.4)^2}{12} = 30.996667$$

Thus, $r^2 = 1 - (1.55637/30.996667) = .9498$.

There is an approximate 95% reduction in the error of prediction obtained by using the least squares equation $\hat{y} = -0.09333 + .91714x$, instead of $\bar{y}$, to predict y.

A.33 a. The **Y** matrix would be $n \times 1 = 18 \times 1$.

b. $$\mathbf{X} = \begin{bmatrix} 1 & 1 \\ 1 & 1 \\ 1 & 1 \\ 1 & 2 \\ 1 & 2 \\ 1 & 2 \\ 1 & 3 \\ 1 & 3 \\ 1 & 3 \\ 1 & 4 \\ 1 & 4 \\ 1 & 4 \\ 1 & 5 \\ 1 & 5 \\ 1 & 5 \\ 1 & 6 \\ 1 & 6 \\ 1 & 6 \end{bmatrix}$$

For each additional replication, we add one additional row for each value of x.

c. The elements of the $\mathbf{X'X}$ matrix for two replications are equal to the elements of the $\mathbf{X'X}$ matrix for a single replication multiplied by 2. The elements of the $\mathbf{X'X}$ matrix for three replications should be the elements of the $\mathbf{X'X}$ for a single replication multiplied by 3. In general, for k replications, the $\mathbf{X'X}$ matrix is k times the $\mathbf{X'X}$ matrix for a single replications.

$$\mathbf{X'X} \text{ (for 3 replications)} = 3\begin{bmatrix} 6 & 21 \\ 21 & 91 \end{bmatrix} = \begin{bmatrix} 18 & 63 \\ 63 & 273 \end{bmatrix}$$

The $(\mathbf{X'X})^{-1}$ matrix for k replications can be obtained by multiplying the $(\mathbf{X'X})^{-1}$ matrix for a single replication by $\frac{1}{k}$.

$$(\mathbf{X'X})^{-1} \text{ (for 3 replications)} = 1/3\begin{bmatrix} 13/15 & -7/35 \\ -7/35 & 2/35 \end{bmatrix} = \begin{bmatrix} 13/45 & -7/105 \\ -7/105 & 2/105 \end{bmatrix}$$

e. $$a'(\mathbf{X'X})^{-1}a \text{ (for 3 replications)} = \begin{bmatrix} 1 & 4.5 \end{bmatrix}\begin{bmatrix} 13/45 & -7/105 \\ -7/105 & 2/105 \end{bmatrix}\begin{bmatrix} 1 \\ 4.5 \end{bmatrix} = .0746$$

$$a'(\mathbf{X'X})^{-1}a \text{ (for 1 replication)} = \begin{bmatrix} 1 & 4.5 \end{bmatrix}\begin{bmatrix} 13/15 & -7/35 \\ -7/35 & 2/35 \end{bmatrix}\begin{bmatrix} 1 \\ 4.5 \end{bmatrix} = .2238$$

Notice that

$$a'(\mathbf{X'X})^{-1}a \text{ (for 3 replications)} = 1/3[a'(\mathbf{X'X})^{-1}a] \text{ (for 1 replication)}$$

In general,

$$a'(\mathbf{X'X})^{-1}a \text{ (for } k \text{ replications)} = 1/k[a'(\mathbf{X'X})^{-1}a] \text{ (for 1 replication)}$$

f. The form of the confidence interval is $\hat{y} \pm t_{\alpha/2}s\sqrt{a'(\mathbf{X'X})^{-1}a}$

If we assume s stays the same, $\alpha = .10$, and 1 replicate has 6 observations, then the following is true:

$$\text{For 2 replicates} \Rightarrow \hat{y} \pm t_{.05,10}s\sqrt{\frac{1}{2}a'(\mathbf{X'X})^{-1}a\text{(for 1 replicate)}}$$

$$\Rightarrow \hat{y} \pm 1.812s\sqrt{\frac{1}{2}}\sqrt{a'(\mathbf{X'X})^{-1}a\text{(for 1 replicate)}}$$

$$\text{For 3 replicates} \Rightarrow \hat{y} \pm t_{.05,16}s\sqrt{\frac{1}{3}a'(\mathbf{X'X})^{-1}a\text{(for 1 replicate)}}$$

$$\Rightarrow \hat{y} \pm 1.746s\sqrt{\frac{1}{3}}\sqrt{a'(\mathbf{X'X})^{-1}a\text{(for 1 replicate)}}$$

The width of the interval for 2 replicates is:

$$2(1.812)s\sqrt{\frac{1}{2}}\sqrt{a'(\mathbf{X}'\mathbf{X})^{-1}a} = 2.56255s\sqrt{a'(\mathbf{X}'\mathbf{X})^{-1}a}$$

The width of the interval for 3 replicates is:

$$2(1.746)s\sqrt{\frac{1}{3}}\sqrt{a'(\mathbf{X}'\mathbf{X})^{-1}a} = 2.01611s\sqrt{a'(\mathbf{X}'\mathbf{X})^{-1}a}$$

Thus, the width of the interval for 3 replicates is only

$$\frac{2.01611\ s\sqrt{a'(\mathbf{X}'\mathbf{X})^{-1}a}}{2.56255\ s\sqrt{a'(\mathbf{X}'\mathbf{X})^{-1}a}} = .787$$

as wide as that for 2 replicates. The width has been reduced by $\approx$ .213.

Appendix B

A Procedure for Inverting a Matrix

B.1 a. $\mathbf{A} = \begin{bmatrix} 3 & 2 \\ 4 & 5 \end{bmatrix} \quad \mathbf{I} = \begin{bmatrix} 1 & 0 \\ 0 & 1 \end{bmatrix}$

Operation 1 Multiply row 1 by 1/3:

$$\rightarrow \begin{bmatrix} 1 & 2/3 \\ 4 & 5 \end{bmatrix} \begin{bmatrix} 1/3 & 0 \\ 0 & 1 \end{bmatrix}$$

Operation 2 Multiply row 1 by 4 and subtract from row 2:

$$\rightarrow \begin{bmatrix} 1 & 2/3 \\ 0 & 7/3 \end{bmatrix} \begin{bmatrix} 1/3 & 0 \\ -4/3 & 1 \end{bmatrix}$$

Operation 3 Multiply row 2 by 3/7:

$$\rightarrow \begin{bmatrix} 1 & 2/3 \\ 0 & 1 \end{bmatrix} \begin{bmatrix} 1/3 & 0 \\ -4/7 & 3/7 \end{bmatrix}$$

Operation 4 Multiply row 2 by 2/3 and subtract from row 1:

$$\rightarrow \begin{bmatrix} 1 & 0 \\ 0 & 1 \end{bmatrix} \begin{bmatrix} 5/7 & -2/7 \\ -4/7 & 3/7 \end{bmatrix}$$

Thus, $\mathbf{A}^{-1} = \begin{bmatrix} 5/7 & -2/7 \\ -4/7 & 3/7 \end{bmatrix}$

Notice: $\mathbf{A}^{-1}\mathbf{A} = \begin{bmatrix} 5/7 & -2/7 \\ -4/7 & 3/7 \end{bmatrix} \begin{bmatrix} 3 & 2 \\ 4 & 5 \end{bmatrix} = \begin{bmatrix} 1 & 0 \\ 0 & 1 \end{bmatrix}$

b. $\mathbf{A} = \begin{bmatrix} 3 & 0 & -2 \\ 1 & 4 & 2 \\ 5 & 1 & 1 \end{bmatrix} \quad \mathbf{I} = \begin{bmatrix} 1 & 0 & 0 \\ 0 & 1 & 0 \\ 0 & 0 & 1 \end{bmatrix}$

Operation 1 Multiply row 3 by 3 and subtract from it row 1 multiplied by 5.

$$\begin{bmatrix} 3 & 0 & -2 \\ 1 & 4 & 2 \\ 0 & 3 & 13 \end{bmatrix} \begin{bmatrix} 1 & 0 & 0 \\ 0 & 1 & 0 \\ -5 & 0 & 3 \end{bmatrix}.$$

Operation 2 Multiply row 2 by 3 and subtract row 1 from it.

$$\begin{bmatrix} 3 & 0 & -2 \\ 0 & 12 & 8 \\ 0 & 3 & 13 \end{bmatrix} \begin{bmatrix} 1 & 0 & 0 \\ -1 & 3 & 0 \\ -5 & 0 & 3 \end{bmatrix}$$

Operation 3 Multiply row 3 by 4 and subtract row 2 from it.

$$\begin{bmatrix} 3 & 0 & -2 \\ 0 & 12 & 8 \\ 0 & 0 & 44 \end{bmatrix} \begin{bmatrix} 1 & 0 & 0 \\ -1 & 3 & 0 \\ -19 & -3 & 12 \end{bmatrix}$$

Operation 4 Multiply row 2 by 11 and subtract from it row 3 multiplied by 2.

$$\begin{bmatrix} 3 & 0 & -2 \\ 0 & 132 & 0 \\ 0 & 0 & 44 \end{bmatrix} \begin{bmatrix} 1 & 0 & 0 \\ 27 & 39 & -24 \\ -19 & -3 & 12 \end{bmatrix}$$

Operation 5 Multiply row 1 by 22 and add row 3 to it.

$$\begin{bmatrix} 66 & 0 & 0 \\ 0 & 132 & 0 \\ 0 & 0 & 44 \end{bmatrix} \begin{bmatrix} 3 & -3 & 12 \\ 27 & 39 & -24 \\ -19 & -3 & 12 \end{bmatrix}$$

Operation 6 Divide row 1 by 66.

$$\begin{bmatrix} 1 & 0 & 0 \\ 0 & 132 & 0 \\ 0 & 0 & 44 \end{bmatrix} \begin{bmatrix} \frac{1}{22} & -\frac{1}{22} & \frac{2}{11} \\ 27 & 39 & -24 \\ -19 & -3 & 12 \end{bmatrix}$$

Operation 7 Divide row 2 by 132.

$$\begin{bmatrix} 1 & 0 & 0 \\ 0 & 1 & 0 \\ 0 & 0 & 44 \end{bmatrix} \begin{bmatrix} \frac{1}{22} & -\frac{1}{22} & \frac{2}{11} \\ \frac{9}{44} & \frac{13}{44} & -\frac{2}{11} \\ -19 & -3 & 12 \end{bmatrix}$$

Operation 8 Divide row 3 by 44.

$$\begin{bmatrix} 1 & 0 & 0 \\ 0 & 1 & 0 \\ 0 & 0 & 1 \end{bmatrix} \begin{bmatrix} \frac{1}{22} & -\frac{1}{22} & \frac{2}{11} \\ \frac{9}{44} & \frac{13}{44} & -\frac{2}{11} \\ -\frac{19}{44} & -\frac{3}{44} & \frac{3}{11} \end{bmatrix}$$

$$\text{Thus, } \mathbf{A}^{-1} = \begin{bmatrix} \frac{1}{22} & -\frac{1}{22} & \frac{2}{11} \\ \frac{9}{44} & \frac{13}{44} & -\frac{2}{11} \\ -\frac{19}{44} & -\frac{3}{44} & \frac{3}{11} \end{bmatrix}$$

c. $\mathbf{A} = \begin{bmatrix} 1 & 0 & 1 \\ 0 & 2 & 1 \\ 1 & 1 & 3 \end{bmatrix} \qquad \mathbf{I} = \begin{bmatrix} 1 & 0 & 0 \\ 0 & 1 & 0 \\ 0 & 0 & 1 \end{bmatrix}$

Operation 1 Subtract row 1 from row 3:

$$\begin{bmatrix} 1 & 0 & 1 \\ 0 & 2 & 1 \\ 0 & 1 & 2 \end{bmatrix} \begin{bmatrix} 1 & 0 & 0 \\ 0 & 1 & 0 \\ -1 & 0 & 1 \end{bmatrix}$$

Operation 2 Multiply row 2 by 1/2:

$$\begin{bmatrix} 1 & 0 & 1 \\ 0 & 1 & 1/2 \\ 0 & 1 & 2 \end{bmatrix} \begin{bmatrix} 1 & 0 & 0 \\ 0 & 1/2 & 0 \\ -1 & 0 & 1 \end{bmatrix}$$

Operation 3 Subtract row 2 from row 3:

$$\begin{bmatrix} 1 & 0 & 1 \\ 0 & 1 & 1/2 \\ 0 & 0 & 3/2 \end{bmatrix} \begin{bmatrix} 1 & 0 & 0 \\ 0 & 1/2 & 0 \\ -1 & -1/2 & 1 \end{bmatrix}$$

Operation 4 Multiply row 3 by 2/3:

$$\begin{bmatrix} 1 & 0 & 1 \\ 0 & 1 & 1/2 \\ 0 & 0 & 1 \end{bmatrix} \begin{bmatrix} 1 & 0 & 0 \\ 0 & 1/2 & 0 \\ -2/3 & -1/3 & 2/3 \end{bmatrix}$$

Operation 5 Operate on row 2 by subtracting 1/2 of row 3:

$$\begin{bmatrix} 1 & 0 & 1 \\ 0 & 1 & 0 \\ 0 & 0 & 1 \end{bmatrix} \begin{bmatrix} 1 & 0 & 0 \\ 1/3 & 2/3 & -1/3 \\ -2/3 & -1/3 & 2/3 \end{bmatrix}$$

Operation 6 Operate on row 1 by subtracting row 3:

$$\begin{bmatrix} 1 & 0 & 0 \\ 0 & 1 & 0 \\ 0 & 0 & 1 \end{bmatrix} \begin{bmatrix} 5/3 & 1/3 & -2/3 \\ 1/3 & 2/3 & -1/3 \\ -2/3 & -1/3 & 2/3 \end{bmatrix}$$

Thus, $\mathbf{A}^{-1} = \begin{bmatrix} 5/3 & 1/3 & -2/3 \\ 1/3 & 2/3 & -1/3 \\ -2/3 & -1/3 & 2/3 \end{bmatrix}$

Notice: $\mathbf{A}^{-1}\mathbf{A} = \begin{bmatrix} 5/3 & 1/3 & -2/3 \\ 1/3 & 2/3 & -1/3 \\ -2/3 & -1/3 & 2/3 \end{bmatrix} \begin{bmatrix} 1 & 0 & 1 \\ 0 & 2 & 1 \\ 1 & 1 & 3 \end{bmatrix} = \begin{bmatrix} 1 & 0 & 0 \\ 0 & 1 & 0 \\ 0 & 0 & 1 \end{bmatrix}$

d. $\mathbf{A} = \begin{bmatrix} 4 & 0 & 10 \\ 0 & 10 & 0 \\ 10 & 0 & 5 \end{bmatrix} \quad \mathbf{I} = \begin{bmatrix} 1 & 0 & 0 \\ 0 & 1 & 0 \\ 0 & 0 & 1 \end{bmatrix}$

Operation 1 Multiply row 1 by 10/4 and then subtract it from row 3.

$$\begin{bmatrix} 24 & 0 & 10 \\ 0 & 10 & 0 \\ 0 & 0 & -20 \end{bmatrix} \begin{bmatrix} 1 & 0 & 0 \\ 0 & 1 & 0 \\ -10/4 & 0 & 1 \end{bmatrix}$$

Operation 2 Divide row 3 by 2 and then add it to row 1.

$$\begin{bmatrix} 4 & 0 & 0 \\ 0 & 10 & 0 \\ 0 & 0 & -20 \end{bmatrix} \begin{bmatrix} -1/4 & 0 & 1/2 \\ 0 & 1 & 0 \\ -10/4 & 0 & 1 \end{bmatrix}$$

Operation 3 Divide row 1 by 4, row 2 by 10, and row 3 by -20.

$$\begin{bmatrix} 1 & 0 & 0 \\ 0 & 1 & 0 \\ 0 & 0 & 1 \end{bmatrix} \begin{bmatrix} -1/16 & 0 & 1/8 \\ 0 & 1/10 & 0 \\ -1/8 & 0 & -1/20 \end{bmatrix}$$

Thus, $\mathbf{A}^{-1} = \begin{bmatrix} -1/16 & 0 & 1/8 \\ 0 & 1/10 & 0 \\ -1/8 & 0 & -1/20 \end{bmatrix}$